Cambridge Archaeological a

THE PLACE-NAMES

OF

NOTTINGHAMSHIRE

THE PLACE-NAMES

OF

NOTTINGHAMSHIRE

THEIR ORIGIN AND DEVELOPMENT

by

HEINRICH MUTSCHMANN,

M.A. (LIVERPOOL), PH.D. (BONN)

Lecturer in German and in Phonetics at the University College, Nottingham

Cambridge:
at the University Press
1913

CAMBRIDGE UNIVERSITY PRESS
Cambridge, New York, Melbourne, Madrid, Cape Town,
Singapore, São Paulo, Delhi, Tokyo, Mexico City

Cambridge University Press
The Edinburgh Building, Cambridge CB2 8RU, UK

Published in the United States of America by Cambridge University Press, New York

www.cambridge.org
Information on this title: www.cambridge.org/9781107665415

© Cambridge University Press 1913

First published 1913
First paperback edition 2011

A catalogue record for this publication is available from the British Library

ISBN 978-1-107-66541-5 Paperback

PREFACE

FOR a first introduction to the science of Names I am indebted to the late Dr Felix Solmsen's lectures (Über Namen, besonders griechische, lateinische, deutsche), delivered in the University of Bonn, in 1905.

This present work on the Place-Names of Nottinghamshire was originally written as a thesis in the School of English Language and Philology of the University of Liverpool. The subject was suggested to me by Professor H. C. Wyld, to whose teaching and kind assistance I owe much.

The field of place-name research is a distinctly dangerous one, and it was only after long hesitation that I decided to lay this study before the public. At one time I was quite prepared to suppress the work entirely, although much time and energy had been spent on its composition. If now given to the world it is because I have been persuaded that its perusal may afford pleasure and instruction to some, and that the theories—often very bold—propounded in the book may draw valuable comments from its critics. It is also hoped that it will stimulate research in a much neglected province of Germanic philology.

It is my pleasant duty to express my sincerest thanks to all who have assisted me in the writing and printing of this book. Mr Robert Mellors (Author of *In and about Nottinghamshire*, etc.) has throughout placed his great local knowledge at my service. Dr F. J. Curtis, B.A., Ph.D., Professor of English Language and Literature in the *Akademie* of Frankfurt am Main, has with great kindness read through the proofs and suggested many useful emendations.

I must here also express my gratitude to Mr J. Potter Briscoe, the Nottingham City Librarian, for having afforded me every facility for using the volumes of early records and other important works which are in the Reference Library, and also for having obtained for me a number of books indispensable to a writer on philological subjects, but not usually contained in provincial libraries.

To all those by whose generous aid the issue of this book has been made possible I wish to express my grateful thanks. Contributions towards a publishing fund were received from: The Faculty of Arts of the University of Liverpool; The Nottinghamshire Society of London; His Grace the Duke of Portland; The Right Honourable the Earl Manvers; The Lord Bishop of Southwell; Sir Thomas Birkin, Bart.; H. Hampton Copnall, Esq., Clerk of the Peace; Principal Heaton; Jesse Hind, Esq., J.P.; W. H. Mason, Esq., J.P.; Colonel Mellish, D.L.; Robert Mellors, Esq.; Major Robertson, J.P.

H. M.

Weimar, August 1913.

TABLE OF CONTENTS

LIST OF ABBREVIATIONS

N.B. For Abbreviations of Sources of Early Forms see Bibliography, Part I.

Germ.	German.
M.E.	Middle English (c. 1050—c. 1500).
M.H.G.	Middle High German.
O.E.	Old English (or Anglo-Saxon, c. 650—c. 1050).
O.H.G.	Old High German.
O.N.	Old Norse (or Scandinavian).
pers. n. (ns.)	personal name(s).
pl. n. (ns.)	place-name(s).
Scand.	Scandinavian (or Norse).
W. Sax.	West Saxon.

Dial. Dict.	Wright's Dialect Dictionary.
Dial. Gramm.	Wright's Dialect Grammar.
N.E.D.	New English (or Oxford) Dictionary.
Vigf.	Vigfusson's Icelandic Dictionary.

*** For full titles of the above and other Works of Reference see Bibliography.

An asterisk (*) before a word denotes a reconstructed or hypothetical form.

A query (?) denotes a doubtful etymology.

> ...develops into....

< ...is derived from....

TABLE OF PHONETIC SYMBOLS

Vowels				*Consonants*		
[æ]	as in	b*a*t		[þ]	as in	*th*in
[ā]	„	f*a*ther		[ð]	„	*th*is
[e]	„	b*e*t		[r]	„	*r*od, a*rr*ow
[i]	„	b*i*t		[s]	„	*s*ee, pla*c*e
[ī]	„	b*ea*t		[z]	„	si*z*e, ri*s*e
[o]	„	p*o*t		[ʃ]	„	fi*sh*
[ɔ]	„	l*aw*		[ʒ]	„	rou*g*e, he*dg*e [hedʒ]
[u]	„	p*u*t		[j]	„	*y*ear
[ū]	„	b*oo*t		[ŋ]	„	si*ng*
[a]	„	c*u*t		[x]	„	German do*ch*, Scotch
[ʌ̄]	„	b*i*rd				lo*ch*
[ə]	„	fath*er*, sof*a*				
[æə]	„	c*are*				
[ai]	„	n*i*ne				
[au]	„	h*ou*se				
[ei, ē]	„	r*ei*n, l*a*ne				
[oi]	„	b*oi*l				
[ou, ō]	„	l*ow*, b*o*ne				

⁂ The other Consonant Symbols have their usual values.

Stress is marked thus [′ɔ].
Phonetic representations are usually placed within square brackets.

NOTE. The phonetic forms in square brackets after the names represent the *local* pronunciation. In most cases a *polite* pronunciation closely following the spelling exists, but is not specially recorded.

Transliterations enclosed in round brackets are taken from Hope's *Glossary of Dialectal Place Nomenclature*, 1883.

INTRODUCTION

BRITISH Place-Names that have an obvious meaning such as *Clifton, Red Hill, Horsepool, Newthorpe*, are very few in number, and often of but recent origin. The majority seem at first sight mere arbitrary conglomerations of sounds having no perceptible relation to the localities with which they are associated. The names *Nottingham, Trent, Cropwell*, etc., are in everyday use; we know the places or objects to which they apply—but we do not know why there should be any connection between them. That such must have existed when the name was first given, or rather sprang into being, will hardly be disputed. The exact nature of this connection, or, in other words, the origin and meaning of the place-name, has at all times been a favourite subject of speculation, both to the learned and the ignorant alike. The attempts of the latter class, besides producing popular etymologies, have given rise to many quaint tales and stories, invented to endow with some significance an otherwise obscure name. Thus *Mansfield* is said to derive its name from a count of *Mansfeld*, in Saxony, who is supposed to have taken part in a tournament held in the famous field near by. Similarly, *Styrrup*, in the same district, is held by some to be "in some way or other connected with the training of horses" for purposes of the noble art of tourneying; whereas *Blyth* has the reputation of being named after "the mirth and good-fellowship of the inhabitants therein." Many more such curious items might be adduced if this were the proper place for their recital. We will, however, take leave of this fascinating subject with the mere mention of that ingenious divine who, "by the slightest change in orthography," made most of the village names round Nottingham have some reference to Baal and to high places.

Much more dangerous than these obviously wrong etymologies are those advanced, often with a great show of learning, by devoted amateurs, chiefly antiquarians or geographers. A common

feature of writers of this class is that they imagine it their chief
duty to explain not so much the nature and meaning of the
name, as the reason *why* it was given to the locality. They
approach the question with the particular bias of their favourite
subject, and very often with preconceived ideas. Thus one will
be alert to discover references to prehistoric settlements ; another
is bent on finding the natural features of the neighbourhood
embodied in the nomenclature of the district; a third will
connect the names of places with persons or events belonging to
national and local history.

COMPOSITION OF PLACE-NAMES.

It is a well-known characteristic of the majority of Teutonic
personal and local names that they consist of two elements or
themes. English place-names of one theme only were very few
from the beginning, and popular etymology has since been at
work changing the appearance of these few so as to make them
conform to the majority. Here belong various Old English
names, originally in the dative case, whose ending -*um* came to
be written -*ham*, as if it represented O.E. *hām*, " home," as *e.g.* in
Askham, Averham, Kelham, Laneham. Lound, Clumber, Coates,
are also examples of uncompounded place-names.

Bi-thematic names almost invariably contain as their second
element a noun of a descriptive character, denoting either a
natural object, such as *wood, field, stone, cliff*; or a work of man,
such as *ton* (" town "), *worth, thorpe, borough.* The first element,
which has also been described as the adjectival theme, is of a
different, a qualifying character. It may consist of an adjective
proper, as in *Radcliffe* (" red "), *Cuckney* (" quick "); of a common
noun used adjectivally, as in *Flintham, Stapleford.* But by far
the most frequent mode of forming English, and for that matter,
Teutonic place-names is by prefixing a personal name descriptive
of the original settler, the owner, inhabitant, or other person
connected with the locality. Very often the personal name
involved does not appear in the singular, but in the plural of a
collective patronymic ending in -*ing*, and meaning " the family,
or descendants of so and so." The persons whose names are
thus perpetuated are almost without exception unknown to

history; no doubt they were often but simple peasants, cottagers, or even serfs. Place-names of this kind, therefore, fail to appeal to the imagination; they are sadly lacking in romance. The only good that can be said in their favour ·is that they have served, and are still serving, an excellent purpose in practical life, and that they provide with amusement the philologist whose business and delight it is to explain the changes which they have undergone in their passage through the centuries.

PLACE-NAME RESEARCH A LINGUISTIC PROBLEM.

Place-names are *words* in the first instance, and as such their elucidation is primarily a linguistic problem. This will become clear on examining the causes that make place-names unintelligible. These causes are manifold, but the most important may be tabulated as follows:

(1) Certain elements contained and preserved in place-names have disappeared from everyday language; *e.g.* -*by*; -*bourne*; and the majority of old personal names.

(2) Certain elements have, in their independent form, assumed a new meaning; *e.g.* -*ton* (= *town*); *well*; *beast* (in *Bestwood*).

(3) Old genuine dialect words (or forms) have become obsolete, because ousted and superseded by forms of the literary language; *e.g. cuck* (= *quick*) in *Cuckney*.

(4) The place-name may be derived from a foreign tongue; *e.g. Trent, Doverbeck*.

(5) The elements contained in a place-name follow developments different from those of the independent words. In a composite name they are more subject to the simplifying processes of shortening and assimilation. For examples of excessive shortening see *Broxtow, Bassetlaw, Caunton*. See also special chapter on Assimilation (Phonology, § 13).

(6) Popular etymology often obscures the original meaning of place-names. See *Arnold, Askham, Birkland, Cropwell, Eastwood, Hempshill, Kingston, Martin, Oldcoates*, etc.

(7) The influence of the spelling interferes with the natural development of place-names. The prevailing and inevitable tendency is to pronounce the names " as they are spelt," although

the written form is very often no sure guide to the etymology. Thus a number of place-names whose first element was a Norse personal name containing the adjectival theme *Thor-*, are now pronounced with initial *t*, because *t* was written by Anglo-Norman scribes for the sound of *th*.

METHODS OF INVESTIGATION.

From the foregoing exposition it is clearly evident that the investigator of place-names cannot base his theories on the modern forms which are the result of the change and wear of centuries. It is necessary to go back to the oldest available spellings, which have to be laboriously collected from a variety of documents printed, for the most part, in the invaluable series of official government publications. Owing to the County's position away from the centre of West Saxon rule, the number of Old English charters relating to Nottinghamshire is exceedingly small. The few documents of this description to be found in Kemble's *Codex Diplomaticus*, and Birch's *Cartularium Saxonicum*, are, moreover, very unreliable, and probably late copies or forgeries. The County is but poorly represented in *Doomsday Book*, and the number of local records of a civil and monastic nature, which are available in published form is lamentably small. This is the more to be regretted as local documents very often contain much more useful spellings than the national ones, which were, particularly in the reigns immediately following the Conquest, often drawn up by Norman-French scribes or by other persons unacquainted with the localities and the speech habits of the people.

Having collected as many early spellings as possible, the investigator proceeds to arrange them in chronological order. Thus the changes that a name has undergone are illustrated, though in a great many cases philological explanations of a highly technical nature are necessary in order to reconcile the various spellings with each other. Often it will be found necessary to distinguish various *Types*, each of which should be treated separately. These owe their existence to a variety of causes. A place-name is sometimes found both in the nominative and dative cases; substitution of elements occasionally occurs as the result of popular etymology; the first element, if a personal name, may have the genitival ending or not; one type may represent the local pronunciation as distinct from the official spelling, etc., etc. These are a few examples of the causes leading to the development of different types.

Each of these types has its own history; but only one can be represented by any given modern form. As a rule, all but one are gradually eliminated; yet it sometimes happens that a local or old-fashioned pronunciation is descended from a different type from the one which survives in the modern spelling.

In this connection, particular attention should be paid to possible errors committed by the Norman-French scribes of *Doomsday Book* and other feudal records. Many of the genuine English sounds were unfamiliar to them, and they often blundered in the rendering of native English words, or modified the pronunciation, and consequently, the spelling of the place-names according to their own French speech habits (see special chapters on Norman Influence, Phonology, §§ 11, 22).

It is, therefore, the philologist on whom devolves the duty of elucidating the meaning and the history of place-names. In his task he should be aided by the local topographer and antiquarian; but as there is, unfortunately, no organisation to ensure the co-operation of all classes of investigators concerned, the philologist will occasionally go wrong for want of local knowledge. Thus, for instance, the local pronunciation of place-names is often as valuable as a very old and genuine spelling; and yet it is nothing short of impossible for the individual and isolated worker to collect a complete and reliable list of such pronunciations.

There is one further aspect of place-name research to which attention must be called. The investigator should not confine himself to the contemplation of the names of one single district. He should go further afield, and study as far as possible the principles and peculiarities of English, and also Teutonic nomenclature generally. Outside England, those countries from which the majority of settlers were drawn, Low Germany, and, in a slightly less degree, Scandinavia, will supply useful analogies.

Some Characteristics of Nottinghamshire Place-Names.

A cursory examination of the main part of this book will convince the reader that the bulk of the place-names of Nottinghamshire are of Anglo-Saxon origin. A considerable number of Scandinavian elements are present; and if these were marked in a distinctive colour on a map, very interesting conclusions might be drawn from the nature of their distribution. I believe they would show that the Scandinavian invasion, which ultimately led to the settlement of the Norsemen in large numbers, was of a comparatively peaceful nature. The Northern newcomers apparently did not try to oust the original occupants of the land, but were satisfied to settle in the marshy, sandy, unattractive regions left vacant by the Anglo-Saxons.

The British and Roman settlements, which no doubt existed anterior to the advent of the Teutonic invaders, seem to have completely lost their original names. Only a few indications of Roman occupation are left. *Brough* and *Littleborough* refer to pre-Anglo-Saxon structures of Roman origin, whereas Celtic elements survive in the river-names[1] *Trent, Doverbeck, Devon,* and *Dean.* Some Celtic word may also be contained in the first part of *Mansfield.*

Norman-French influence is apparent more in the modification of native names than in the creation of new ones. *Perlethorpe,* containing a French personal name as first element, is a rare example of a post-Conquest formation on the old principle of Teutonic name composition. *Beauvale* and *Belvoir,* marked by a touch of the romantic spirit appertaining to the age of chivalry, and thus notably distinguished from the bulk of commonplace Germanic names, are instances of completely Norman-French formations. On the other hand, distinctive additions to older place-names are frequently of Norman-French origin. They usually take the form of the new feudal owner's name which is tacked on to the older native name of the place, as in *Cropwell Butler, Holme Pierrepont,* etc.

---

[1] The Rev. John Sephton, M.A., in his *Handbook of Lancashire Place-Names* (Liverpool, 1913), expresses the very interesting opinion that the vitality of Celtic river-names is due to religious or superstitious causes (p. 132).

NOTTINGHAMSHIRE PLACE-NAMES

ADBOLTON (in Holme-Pierrepont).

Type I.

1086 Alboltune, D.B.

Type II.

1316 Adbolton, Bor. Rec.
1346 Adbolton, F.A.
1571 Adbolton, Index.

" The *tūn* or farmstead of *Ealdbeald* (Type I), or *Eādbeald*
(Type II)." Both personal names occur frequently. Interchange
of prefixes is found in the pers. ns. themselves : *Ealdbeald*, king
of Kent, appears as *Eādbeald* in some sources (see Onomasticon).

ALVERTON [olvətn].

1086 Alvretun, Alvritun, D.B.
1278 ⎫
1304 ⎬ Alverton
1316 ⎭
⎧ H.R.
⎨ Index.
⎩ F.A.

" The *tūn* of *Alfer*," O.E. *Ælfherestūn* ; *Alfer* is an O.E. and
M.E. short form of *Ælfhere* (Onomasticon).

ALWOLDESTORP (not identified).

1086 Alwoldestorp, D.B.

" The *þorp* of *Ealdweald*."

M.

ANNESLEY [ænzli].

1086 Aneslei, D.B.
1240 Anyslegh, Bor. Rec.
1284 Anisley, F.A.
1421 Annesley, Index.

"The *lēah* (lea) of *Anna*." *Anna* is an O.E. man's name. The early substitution of the strong for the weak declension is characteristic of the Northern and Midland dialects; Sweet, N. Engl. Gramm. § 989, Sievers, § 276 *a* 5; see Alexander, Mod. Language Rev. 1911.

APPESTHORPE or HABBLESTHORPE.

Type I.
1278 Harpelestorp, H.R.
1327–77 Harplesthorp, Non. Inq.

Type II.
1278 Happelestorp, H.R.
1316 Apullesthorp, F.A.
c. 1500 $\begin{cases} \text{Apulthorp} \\ \text{Habylthorp} \end{cases}$ Inq. P.M. c. 1500.

"The *þorp* of * *Harpel* or * *Arpel*." The *r* of Type I was lost through either assimilation or dissimilation,

$$rp > pp, \text{ or } r - l - r > [] - l - r.$$

After this change the name became connected with *apple*. I cannot trace the pers. n. (*H*)*arpel* in other English sources. It is, however, found in continental records. It appears to be the diminutive of *Arpus*, which is the name of a German chief mentioned by Tacitus, Ann. II, 7. See Much, Zeitschrift für deutsches Altertum, 35, 365, Grimm, Geschichte der deutschen Sprache, 580. The pl. n. *Erpelingalanda* is recorded by Förstemann, II².

ARNOLD.

1086 Ernehale, D.B.
1157 Ric. de Erneshala, P.R.
1221 Arnehale, Bor. Rec.
1284 Arnale, F.A.

1272–1307 Arnehal, Index.

1316 Arnall, F.A.

1346 Arnale, F.A.

1428 $\begin{cases} \text{Arnewall} \\ \text{Arnal} \end{cases}$ F.A.

c. 1500 $\begin{cases} \text{Arnall} \\ \text{Arnell} \end{cases}$ Inq. P.M. c. 1500.

"The *healh* of *Earne*." Although the genitival *s* is encountered but once, the first element can hardly represent the O.E. *earn*, "eagle." The final *d* is excrescent; similar cases of the development of *d* are found in various other words both in the dialects and in literary speech; see Wright, Dial. Gramm. § 306; Horn, Neuengl. Gramm. § 188.

ASKHAM.

1086 Ascam, D.B.

1278 Ascam, H.R.

O.E. *æt æscum*, "at the ash-trees," a regular dative plural. The modern spelling is due to popular etymology: the ending *ham* in the modern form of this name has nothing to do with *home*.

ASLOCKTON or ASLACTON.

1086 $\begin{cases} \text{Aslachetone} \\ \text{Haslachestone} \end{cases}$ D.B.

1302 Aslacton, F.A.

"The *tūn* of *Aslac*." The pers. n. is of Scandinavian origin; see Björkman.

ATTENBOROUGH.

c. 1200 Adigburc $\rbrace$ Woll. MSS.

c. 1240 Hadinbur $\rbrace$

1275 Adinburks, Bor. Rec.

1291 Addingburg, Tax. Eccl.

1327–77 Adyngburgh, Non. Inq.

c. 1500 Addyngborough, Inq. P.M. c. 1500.

Either "the *burh* of the *Eādings* or of *Eadda*." The change from *d* to *t* must be quite recent, and is perhaps due to dissimilation. Similar changes in pers. names are discussed by Bardsley, Dict. of Engl. and Welsh Surnames, p. 19.

AVERHAM [æərəm] (Airham, Hope).

Type I.

1086 Aigrun, D.B.
c. 1200 Egrum, Index.
1278 Egrom, H.R.
1291 Egrum, Tax. Eccl.
1302 Aghram, F.A.

Type II.

1316 Averam, F.A.
1327–77 Averham, Inq. Non.
1637 {Averham (or Aram), Camden, p. 549.
{Havorham, Map in Camden.
1680 Averham, Index.

Type III.

1637 Aram, Camden, p. 549.

*** c. 1600 (?)…Averham, auntiently called Egrum but now comonlie called Aram…MS. BM. Titus A. xxiv. fol. 130 b.

I take this name to represent O.E. (Mercian) *æt æðrum*, "at the waters, streams"; it would thus correspond to the Latin *ad aquas*. The exact meaning of O.E. (W. Saxon) *ǣdre*, Anglian *(h)ēþir, ēþre* is "a channel for liquids, an artery, vein, fountain, river," Bosworth-Toller; the cognate German word *Ader* has the same meaning. The following quotation from White's Directory (1853) will explain the origin of the name: "The large island formed by the two branches of the Trent navigation opposite to Newark, is in the manor of Averham, or Aram…" (p. 460).

I have not yet succeeded in finding another instance of the occurrence of O.E. *ǣdre* in pl. names. There are, however, a few continental names which contain its O.H.G. cognate. Mod. Germ. *Brunnadern* near Bondorf in Baden goes back to an O.H.G. *Brunnaderon*, a dat. plural; see Förstemann, II², 10. According to Graff, Althochdeutscher Sprachschatz, I, 157, *brunadara* is used by Notker in his translation of the Psalms to render the Latin *manationes aquarum*. The same element may be contained in the Hessian river-name *Itterbach*, Sturmfels, Ortsnamen Hessens, p. 41.

The interpretation of the various spellings of this name is not without its difficulties. I shall now endeavour to reconcile the three types with my assumption.—Type I: In Anglo-French records the open *g* sound is frequently substituted for the English *ð*; see Zachrisson, pp. 101, 117; cp. the various spellings of " Leicestershire, Worcester(shire) " in different MSS. of Bede, as *Lepecæstrescire, Lægreceasterscire*; *Wiðreceasterscir, Wigraceaster*, Miller, p. 46.—Type II: On the other hand, intervocalic *ð* often developed into *v* in the English dialects which accounts for the second collection of forms; see Horn, Hist. Neuengl. Grammat. § 197; Wright, Dialect of Windhill, p. 91 : "Fifty years ago, *f* for *p* and *v* for *t* were quite general."—Type III: The pronunciation is [æərəm], with the regular loss of *ð* or *v* before *r* in a medial position; Horn, l.c. § 169.

The development of the various forms may be tabulated as follows :

O.E. (*æt*) *æðrum*.

> *Egrum* in Anglo-French spelling ;

> *Averum* in the local dialect ;

> *Arum* in subsequent local development. The modern spelling in -*ham* is due to confusion with the frequent termination O.E. -*hām*.

N.B. Isaac Taylor's assumption (Words and Places, ch. XI) that *Averham* is derived from the dat. pl. of O.E. *hearg*, " a heathen temple," is untenable.

AWKLEY or AUKLEY.

   1278 Alkelaye, H.R.
   1316 Alkeleye, F.A.
   c. 1500 Aulkeley, Inq. P.M. c. 1500.

"The *lēah* of a man called *Ealca*, or of *Ealce*," a mythological person, or deity. In O.E. the pers. name *Alca* occurs once; it most probably represents a short form of one of the numerous "full-names" beginning with *Ealh-*, *Ealc-*; cf. also *Ealac, Alac*, Onomasticon, and *Alako*, Förstemann, I.

As to the second suggestion it cannot be denied that we find traces of a mythological person of the name of *Ealce* etc.; see

Middendorf, s.v. On Low German territory, in the neighbourhood of Osnabrück, the geographical names *Alke Krug* and *Alk Pool* are found close to an ancient heathen place of worship (Mitteilungen des Vereins für Geschichte und Landeskunde von Osnabrück, XIII, 1886, pp. 263 sqq.). The same deity or deities seem to be mentioned by Tacitus in the Germania, c. 43 : "Apud Naharvalos antiquae religionis lucus ostenditur. praesidet sacerdos muliebri ornatu, sed deos interpretatione Romana Castorem Pollucemque memorant. ea vis numini, nomen *Alcis* (var. *Alces* vel *Alci*). nulla simulacra, nullum peregrinae superstitionis vestigium ; ut fratres tamen, ut iuvenes venerantur." As commentators fail to give a satisfactory explanation of this singular passage, I thought it worth while to quote it in connection with the pl. n. under discussion, hoping that further inquiry will either strengthen or disprove the theory advanced. It is highly interesting to note that these *Alces* were worshipped in a "lucus," which word is closely related to the O.E. *lēah* (see List of Elements, s.v.). The Roman interpretation is not to be implicitly trusted.

The following pl. ns. seem to contain the same first element as *Awkley* :

Alkenthyt, Alkentheyt Hill, Bor. Rec. I, p. 375.

Alkenthweyt, ib. p. 391.

Alchenfluh, in Switzerland (?); *Fluh*, O.H.G. *fluoh*, means "Felswand, Felsabsturz."

AWSWORTH.

1086	Eldesworde, D.B.
1295	Aldesword, Woll. MSS.
1302	Aldisword ⎫ F.A.
1316	Aldesworthe ⎭

"The *weorþ* or homestead, farm of *Ealda*." The O.E. name *Ealda* is of frequent occurrence. It either means "the old one," or more likely is a short form of one of the numerous compound names beginning with *Eald-*, as *Ealdhelm, -here, -weald*. The appearance of an *-s* in an originally weak noun is by no means without parallel; see *Annesley*.

BABBINGTON.

Owing to the absence of early forms, it is impossible to explain this name accurately. The first element is no doubt the O.E. pers. n. *Babba*, which may have appeared either in the weak gen. sing. (O.E. *Babbantūn*), or in the patronymic form (O.E. *Babbingatūn*). The latter forms the first element in the O.E. pl. n. *Babbingdon* (Birch, Cartul. Sax. p. 316), and also in the continental names O.H.G. *Papinga*, *Pappingen* (modern *Pabing*) and *Papingohuson* (Förstemann, II).

BABWORTH.

      1086 Baburde, D.B.

      1316 Babbeworth, F.A.

      1637 Badworth, Map in Camden.

"*Babba's weorþ* or homestead." *Babba* is an O.E. man's name. Camden does not seem to have entertained a very high opinion of the locality.

BAGGALEE (under Greasley).

This may be "*Bagga's lēah*"; but there are no early spellings to support this or any other view. The place is popularly known as *Beggarlee*; can this be the correct etymology? There is a *Beggar's Bush* in Staffordshire, see Duignan, Place-Names of Staffordshire.

BAGTHORPE (under Selston).

"The *þorp* of *Bagga*"? There are no early forms.

BALDERTON.

      1086 Baldretune, D.B.

      1291 } Baldirton {Tax. Eccl.
      1316 }   {F.A.

"The *tūn* of *Bealdhere*"; the latter is an O.E. man's name of which five bearers are known (Onomasticon). The pl. n. has, of course, nothing to do with *Baldr* the Norse deity.

BARNBY (MOOR).

      1086 Barnebi, D.B.

      1445 Barnby Moor, Index.

      1637 Barmbye on the Moor, Map in Camden.

"The *bӯr* or habitation of *Barn*." The suffix clearly shows that the place was a Danish settlement. The pers. n. *Barn* is recorded by Björkman. The *m* in Camden is either due to assimilation, or represents one of the numerous mistakes of the engraver. The district round this place formerly was wild moorland which accounts for the distinctive addition. May we conclude from this fact that when the Danes arrived in this district, they found the best part of the country occupied by the Saxons and had to content themselves with the less alluring portions?

BARNBY-IN-THE-WILLOWS.

   1086 Barnebi, D.B.
   1302 Barneby, F.A.
   1637 Barmby, Map in Camden.

See preceding name. The place is situated on the river *Witham* (q. v.) which derives its name from the numerous willows growing on its banks. The same natural phenomenon supplied this pl. n. with its distinguishing epithet.

BARNSTON.

   1086 Bernestune, D.B.
   ${1286 \atop 1302}$ Berneston ${\text{Index.} \atop \text{F.A.}}$
   1347 Barnstone, Index.
   1637 Burnston, Map in Camden.

"The *tūn* of *Beorn*." The pers. n. *Beorn* is found both in O.E. and in Scandinavian. In the latter language it was particularly frequent; see Björkman.

Camden's spelling represents a different development of the M.E. *er*; this combination either changed into *ar*, or remained unaltered. In the latter case, it coincided with *ir* and *ur* in pronunciation during the 17th century; see Phonology, § 8.

BARTON-IN-FABIS.

   1086 Bartone, D.B.
   1302 ${\text{Berton} \atop \text{Barton}}$ F.A.
   1637 Borton, Map in Camden.

The pl. n. *Barton* is very widely disseminated all over the country. It is usually taken to represent O.E. *bere-tūn*, "corn-farm,

grange," or more literally "barley-enclosure, rick-yard"; see
Lancashire Place-Names, p. 290. It is strange, however, that the
D.B. form should exhibit *ar* instead of *er*; there must have
existed an O.E. *bærlīc*, the ancestor of modern *barley*, which may
have influenced *bere-tūn*, changing the *e* into *æ*; cp. M.E. *barlic*,
Morsbach, M.E. Gramm. § 108, anm. 1, 3. The regular change of
er > *ar* did not take place till the first Modern English period;
Sweet, N.E. Gramm. § 845.

The *o* in Camden's form is due to the rounding influence of
the initial *b*; cp. *Borwick* < **Barwick*, Lancashire Place-Names,
p. 74.

In the middle ages, distinguishing additions to pl. ns. were
often translated into Latin, the language of the documents, as
here *in Fabis*; see Zachrisson, Latin Influence etc. p. 74. The
Leicestershire Barton-in-the-Beans exhibits the same addition
in the native idiom.

BASFORD [beisfəd] (Bāysfud, Hope).

> 1086 Baseford, D.B.
> 1284 Baseford ⎱ F.A.
> 1302 Besseford ⎰
> 1369⎱ Baseford, Index.
> 1412⎰

"The ford of *Bass* or *Bassa*, the ford near which *Bassa* lived."
The *s*, being voiceless in the modern pronunciation, must represent
O.E. *ss*. As the vowel was long in M.E., the lengthening cannot
be due to its standing in an open syllable. We are, therefore,
forced to assume that a lengthening of *a* (or *æ*) took place before
s(s) in early M.E., similar to that of *æ* before *s*, *þ*, *f* in the
18th century (Sweet, N.E. Grammar, § 844; Horn, Histor.
Neuengl. Gramm. § 47, dates this change much earlier). The
F.A. spelling of 1302 shows that by that time the lengthened sound
had been considerably advanced towards the front position.

BASSETLAW (Wapentake, now a Parliamentary Division).

Type I.

> 1155 Desetelawahdr ⎱ P.R.
> 1189 Bersetelaw Wap.⎰
> 1278 Bersetelawe, H.R.

Type II.

1086 Bernedeselawe Wapentac, D.B.

O.E. *bearu-sǣtena-hlǣw*, "the mound of the forest-dwellers";
from O.E. *bearu*, "a wood, *nemus vel lucus*," *sǣta*, "resident, inhabitant," found in compounds only and mostly in the plural as
in the present name (cp. O.E. *dom-, dūn-, burh-, land-sǣtan*,
O.H.G. *himil-sâzo*, "inmate of heaven"). O.E. *hlǣw* may denote
either "an artificial or natural mound." It was the custom of
the men of a hundred and especially of a Scandinavian wapentake
to assemble on a hillock which gave the name to the division.
This mound was sometimes raised artificially, which is the case
of the most remarkable of these *lowes*, Tynwald Hill in the Isle
of Man. (Cp. Binghameshou Wap. in D.B.)

The initial *D* in the P.R. of 1155 is a scribal error; the D.B.
compiler probably imagined the first element of the name to be
derived from the pers. n. *Beornheard* : the long word was too much
for the French clerk.

It is interesting to note that the German *Holstein* has a
similar origin. The name has nothing to do with *Stein*, "stone";
-stein is a popular corruption of *-sten* which is still found
in the name of the *Holstentor* at Lübeck. The Old Saxon
form of the name of the inhabitants is *Holtsâti*, "the dwellers
in the *holt* or wood." The geographical name is, like the
majority of German names of districts, derived from the dat.
plural of the name of the inhabitants. The explanation of
the name given by a mediæval writer and quoted by Förstemann
(II, p. 866) might be applied to the *Bearusǣtan* with equal force:
"Holcete dicti a silvis quas incolunt." White (Directory, 1853,
p. 577) remarks that "the ancient forest of Sherwood (q.v.)
extended over a large portion of this division [i.e. Bassetlaw
Hundred], nearly the whole of which, during the last century,
has been enclosed and though generally a deep light sandy soil,
now forms a rich agricultural district, scarcely equalled in the
kingdom."

The spelling with *e* instead of *a* seems to indicate that M.E.
a was advanced before *r* at an early period (see Phonology, § 7).
The disappearance of *r* before *s* is a frequent phenomenon (see
Phonology, § 7).

Soon after the Norman Conquest, a noble family of the name of *Basset*[1] is found in this hundred; they evidently take their name from the property owned by them in the division. See *Colston-Basset.*

BASSINGFIELD [bæzinfĭld]?

1086 {Basinfelt / Basingfeld} D.B.
1284 Bassingfeld, F.A.
1571 Basingefeild, Index.

"The field of *Basing*," or, more probably, "of the *Basings*."

BATHLEY [bætli]?

1086 Badeleie, D.B.
1316 Batheleye, F.A.
1452 Bathley, Index.

"The lea of the bath," "the meadow, containing a bathing-place." Cæsar (De bello Gallico, IV, 1) informs us that the Germans were very fond of bathing in the open[2], so that it is very natural that they should have left traces of that habit in pl. ns.   There are numerous such names to be encountered on the continent, as *Wiesbaden*, "baths in the meadow(s)," *Baden* of which there are several, from O.H.G. *az badun*, "at the baths." The English *Bath* appears as *æt Baðum* (dat. pl.) in O.E. records.

The first element may, however, represent the O.E. man's name *Bada* which was later on changed into *Bath* by popular etymology.

The pronunciation recorded above is not well authenticated; in any case it would be difficult to account for the *t.*

BEAUVALE (Priory) [bouveil].

Type I (Latin).

1291 (Conventus de) Bella Valle, Tax. Eccles.
c. 1500 Bellavalle, Inq. P.M. c. 1500.
1535 (Prioratus de) Bella Valle, Valor Eccles.

---

[1] Several members of this family are mentioned in documents relating to Notts. and Leicestershire printed in "Calendar of Documents Preserved in France" (Index).

[2] By the time of Tacitus they seem to have become more averse to this violent practice (Germania, ch. XXII).

Type II (French).

c. 1500 Beauvale, Inq. P.M. c. 1500.

Type III (Phonetic or English).

c. 1500 Bovall [for "Bovale"?], Inq. P.M. c. 1500.

1637 Bonall [*n* for *u* = *v*], Map in Camden.

The etymology of this name is clear. It is, however, doubtful whether priority belongs to the Latin or French type. This is an instance of the comparatively rare purely Norman-French pl. ns. in England. Whereas the names of places of Germanic origin are generally of a most commonplace and strictly "practical" character, these Norman names frequently refer to the beauty of the surroundings: the imperious conquerors were able to pick and choose the site of their dwellings. Similar cases are *Beaulieu, Beauchief, Beaumont, Beauchamp*. (See Bradley, Essays and Studies, I, p. 39.)

It may be mentioned here that the valley in which the ruins of the priory are situated fully deserves the appellation.

BECK.

From Scandinavian *bekk(r)*, "brook."

BECKINGHAM.

> 1086 Beching(e)ham, D.B.
> 1189 Bekingeha, P.R.
> 1216–72 Beghenham, Index.
> 1316 Bekyngham, F.A.
> 1637 Beckingham-Supermost, Map in Camden.

"The home of the *Beccings*, the descendants or family of *Becca*." Camden's addition explains itself.

BEESTHORPE.

> 1086 Bestorp, D.B.
> 1204 Bestorp, Index.

The first element of this name may be a pers. n. **Bē* or **Bēs* of which I cannot find reliable traces[1]: there is a *Beesby* in

---

[1] The M.E. name *Bee* (from *Beatrice*) recorded in Bardsley's Dictionary of Engl. and Welsh Surnames is a late formation and cannot be used to explain the form of D.B. It is equally impossible, for obvious reasons, to connect this and the following pl. n. with the female St Bee.

Lincolnshire. It is also possible that *Bees* stands for an old river-name. (See *Beeston.*)

BEESTON [bīsn] (Beesun, Hope).

1086 Bestune, D.B.
c. 1200 Bestona, Woll. MSS.
1284 Beston, F.A.

There may have been an old pers. n. **Bē* or **Bēs* from which the pl. n. is derived. Numerous *Beestons* are found in various parts of England, in Cheshire, Bedford, the West Riding, and Norfolk. Some of these are derived from *Bedestūn*, "the farm of Bede." Dr Moorman (Place-Names of the W. Riding, p. 24) assumes an earlier *Beowestūn* as the origin of the *Beeston* in his district. For this there is, however, no authority.

Considering that many places take their names from the rivers on which they stand (Bradley, Essays and Studies by Members of the English Association, I, p. 32), one might advance the theory that *Beeston* is derived from an old river-name **Bēos-ēa*. This assumption is based on the occurrence of such a name on the continent: a river *Biese* (O.H.G. *Bese*, Förstemann, II) joins the *Aland* in the northern part of the province of Saxony. As many river-names were brought over from the continent by the Anglo-Saxons, this particular one might have been among them (Jellinghaus, Angl. XX, 257 sqq.).

BELVOIR (Vale of) [bīvǝ, belvoiǝ].

1535–43 The vale of *Bever*, baren of Wood, is large and very plentiful of good Corne and Grasse, and lyith in 3. Shires, Leycester, Lincoln, and much in Nottinghamshire
Beavoire (Castelle) ⎬ Leland, I, 108.

1613 *Bever's* batning slade, Drayton, Polyolbion, XXVI, 2.

Although *Belvoir Castel* is situated in Leicestershire, the name is included because the *Vale of B.* lies partly in Notts. The etymology is obvious. Unlike the majority of modern *Bellevues* and similar names, the Castle well deserves its appellation. Cp. *Beauvale.*

BESTHORPE.

1086 Bestorp, D.B.

1302 Besthorp, F.A.

This name seems to be identical in origin with *Beesthorpe* (q.v.), with the vowel shortened before the combination *sþ*.

BESTWOOD (Park).

Type I.

c. 1200 Beescwde, Woll. MSS.

1205 Beswude ⎫
1247 Besekwood⎭ Cal. Rot. Chart.

1535–43 Beskewood, Leland.

1637 Beskwood, Map in Camden.

Type II.

1437⎫
1681⎭ Bestwood Park, Index.

"The enclosed wood where deer are preserved." In meaning, this word corresponds to the O.E. *dēor-frið*. Although the earliest spellings exhibit *k*[1] instead of *t*, I take the latter to be the original letter which was changed to *k* through assimilation (see Phonology, § 13). Thoroton informs us (vol. II, p. 179) that the park was "well stored with deer before the troubles" (i.e. the Civil War?). White's Directory (1885) contains a note to the effect that in 1251 Bestwood was "a Hay or Park of our Lord the King wherein no man commons."

BEVERCOATES.

1302 Bevercotes, F.A.

1637 Bircotes, Map in Camden.

O.E. *beofor cotu*, "the beaver cotes or dwellings." In the transition from O.E. to M.E. the distinctions of grammatical gender were completely lost, and the originally neuter *cot* assumed the plural ending *-es* of the masculine nouns (Sweet, N. Engl. Grammar, § 989).

---

[1] The *c* in the spelling of the Woll. MSS. may be a mistake for *t*; the two letters are frequently interchanged by the scribes on account of their almost identical shape.

This name proves that beavers once were not infrequent inhabitants of this island (cf. Taylor, Words and Places, ch. xv). In the neighbourhood of Bevercoates there is an abundance of brooks and springs; originally the country must have been a wild swamp, just the place for beavers to erect their constructions.

Camden's spelling no doubt represents the contemporary pronunciation.

BILBOROUGH.

> 1086 Bileburg(h), D.B.
> 1180 Billeburg, Woll. MSS.
> 1284 Bilburgh, F.A.

" The *burh* or fortified place of *Billa*."

BILBY.

> 1086 Billebi, D.B.
> 1316 Bylby, F.A.

" The *byr* or farmstead of *Billa*." The second element is of Scandinavian origin; but the pers. n. *Billa* may be either O.E. or Scand.

BILHAGH (a wood of Sherwood Forest).

> 1637 Bellow, Map in Camden.

The second element is O.E. *haʒu*, "a fence, a piece of ground enclosed with a fence." As to the meaning of the first element I have no suggestion to offer. Camden's spelling probably represents the contemporary pronunciation.

BILSTHORPE.

> 1086 Bildestorp, D.B.
> 1233 Bilsthorpe, Index.
> 1291 Bildisthorp, Tax. Eccl.
> 1302 Bildesthorp ⎫ F.A.
> 1428 Bilsthorp ⎭

This name probably means " *Bilheardes þorp*." The pers. n. *Bilheard* is recorded once (Onomasticon). The phonetic development—a continual process of elimination—was as follows:

Bilheardesþorp > *Bilrdesþorp* > *Bildesþorp* > *Bilsþorp*. It is, however, possible that the first element was the Scand. pers. n. *Bildi* (more usually *Billi*), recorded by Rygh, Gamle Personnavne, p. 36. The Scand. character of the second element speaks in favour of such a derivation.

BINGHAM.

$$\left. 1086 \left\{ \begin{array}{l} \text{Bingameshou Wap.} \\ \text{Bingehamhou Wap.} \\ \text{Bingehā} \end{array} \right\} \text{ D.B.} \right.$$

$$\left. \begin{array}{c} 1278 \\ 1284 \end{array} \right\} \text{ Bingham } \left\{ \begin{array}{l} \text{H.R.} \\ \text{F.A.} \end{array} \right.$$

1578 Bingham in le Vale, Index.

"The home of the family of *Benning*," O.E. *Benninga hām*. A contracted **Bengham* would become *Bingham* (see Phonology, § 6). In a charter (Cart. Sax. 125) we find the name *Benninga wurth*; there is a *Binningham* in Yorkshire which is spelt *Beningham* in 1303 (Index), and a *Binnington* in the same county which appears as *Benington* in 1555 (Index). The same patronymic is encountered in continental pl. ns., cf. *Binningen* (Förstemann, II). *Hou* in D.B. is derived from O.E. *hōw*, "a hill"; hundreds and wapentakes were frequently called after hills; see *Bassetlaw*.

BIRKLAND (ancient wood of Sherwood Forest).

1278 foresta dni Reg' int[er] Birkelund & Heselund, H.R.

The meaning is clearly "Birch wood." Both elements of this name are of Scandinavian origin. The former is cognate with O. Icelandic *birki-*, "a birch" (in compounds only, Vgf.); the word is not mentioned in Björkman's book on Scandinavian Loan Words. The suffix is the Scand. *lundr*, "a wood," still found, in various forms, in the English dialects, also as an independent pl. n.; see *Lound*. It was changed to *-land* through popular etymology. This new termination may stand for either the common word *land*, "expanse of country," or M.E. *land, laund*, "wild, shrubby, or grassy plain," derived from O. French *lande*.

BLEASBY.

1278 Blesby, H.R.
1302 Bleseby, F.A.

Various explanations may be offered, although none seems conclusive:

(1) " The *bȳr* of the blast, the windy habitation"; from O.E. *blǣs*, "a blowing, blast." The same element seems to occur in the Lancs. pl. n. *Bleasdale*, which, however, is explained differently by Prof. Wyld (cp. *Lowdham*).

(2) There may have existed a pers. n. *Blǣsa* corresponding to the O.H.G. *Blåso* in *Blasindorf* (Förstemann, I).

(3) The first element may contain the name of a river or brook (cp. *Beeston*). We find two rivers called *Blies* in Germany (cp. Förstemann, II, s.v. *Blesa flumen*).

BLIDWORTH.

1086 Blideworde, D.B.
1157 Blieswurda, P.R.
1278 Blytheworth, H.R.
1598 Blodworth, Index.
1637 Bledworth, Map in Camden.

" The *weorþ* or farm of *Blīþa*"? Before *w*, *ð* seems to have become stopped and changed to *d*. The P.R. form shows loss of intervocalic *ð*, pointing back to an O.E. *Blīðes-weorþ*. It is remarkable that the other forms are without the genitival *s*. The pers. n., which is not recorded in the assumed form, seems to be an abbreviated variety of one of the many names beginning with *Blīð-*, as *Blīðhelm, -here, -mund, -weald* etc. (Onomasticon). The two last quotations appear to be no more than fanciful or erroneous spellings.

BLYTH [blaið].

1086 Blide, D.B.
1153 (?) Blie, Index.
1278 {Blid'
          (Prior de) Blida} H.R.
1316 Blid, F.A.
1327–77 {Blida
            Blyth} Non. Inq.

M.   2

There are a considerable number of rivers called *Blyth* in various parts of England. I take the above name to be derived from the river on which the town stands. The place was originally described as *æt (on) þære blīða(n) ēa*, "at (on) the blithe, gently flowing, calm brook." The *e* in the second spelling and the *-a* in the latinised *Blida* above seem to point to the second element having at one time been *ēa*. The word *blithe*, "laetus, suavis, placidus," is a very appropriate epithet for many of the English streams. I was unable to find a brook of that name near the town of *Blyth*; the following extract from Leland's Itinerary (I, 98), however, proves that one of the water-courses in the neighbourhood once bore that name: "There renne to Brookes as I cam into the Toun of *Blith*, the first that I cam over was the Greatter, and cummith thither from the Weste: the other rennith hard by the utter Houses of the Toun; and this, as they told me, was namid *Blith*."

N.B. The O.E. nom. of the name of the river, which preceded that of the town, was *blīðu ēa*, or *sēo blīðe ēa*.

BOLE.

Type I.

1086 Bolun, D.B.

Type II.

1316 Bole, F.A.
1327–77 Bole super Trent, Non. Inq.
1555 Bolle, Index.

It is impossible to say which is the correct etymology of this name. Type I seems to represent the O.E. dat. pl. of *bóld*, "building, dwelling, house" (see Bülbring, § 522); Type II would, in that case, stand for the dat. sing. The *d* after *l* was assimilated at an early period. The course of development would be as follows: *Bolde > Bolle > Boule > Bole*; the modern spelling correctly indicates the pronunciation but not the etymology.

BOLHAM or BOLLAM.

1278⎫
1335⎭ Bolum ⎰H.R.
⎱Index.

O.E. *æt bóldum*, "at the dwellings, houses," see the preceding name. The spelling in -*ham* is due to a misconception as to the nature of the final syllable.

It is not improbable that the "dwellings" referred to were ancient rock-houses of which traces are still to be found. White's Directory (1853) may be quoted here: "The village formerly had numerous *rock-houses* formed by excavations in the shelving rock of red sandstone, but few of these troglodyte dwellings are now inhabited."

BONBUSK.
> 1571 Bonbusk, Index.

"*Bonda's* bush, or coppice"? The name *Bonda* is very frequent in East Scandinavian sources; see Björkman. The M.E. *buske* is also of Scandinavian origin.

BONINGTON.
> 1086 Bonniton, D.B.
> 1291 Bonigton, Tax. Eccl.
> 1327–77 Bonyton, Non. Inq.
> 1346 Bonyngton, F.A.

O.E. *Boninga tūn*, "the homestead of the descendants of *Bona*." A family of the name of *Baningas* is mentioned in the O.E. poem *Wīdsīþ*, line 19. The *a* before a nasal was frequently changed to *o*; Bülbring, § 123.

The same patronymic is encountered in continental pl. ns. Förstemann, I, records the following: *Boningaham*, and *Boningue*, near Calais.

BOTHAMSALL or BOTTOMSALL [locally: boð̄msəl; otherwise: botmsəl] (Bottomsall, Hope).

> *Type I.*
> 1535 Bodv'sell, Val. Eccl.
>
> *Type II.*
> (1) 1086 Bodmescel(d), D.B.
>   1180 Bodemeskil, Index.
> (2) c. 1200 Bodmeshil, Index.
>   1278 Bodmeshill, H.R.

2—2

    (3)   1225 Botmeshil, Index.

    (4)   1302 Bothemeshull ⎫

           1316 Bothemeshul ⎬ F.A.

           1428 Bothomsell ⎭

"The well or spring of *Bodwine* (or **Bodmǣr*?)." From Type II 1, the earliest spellings on record, we clearly gather that the second element was the Scand. *kelda*, "a spring or well." A flowing well is still to be seen in a field in the centre of the village; its water supplies a trough standing in the main road. It is very probably the original spring after which the locality is called.

There exists some doubt as to the exact significance of the first element. The name *Bodwine* is frequently found in O.E. documents, but a trace of the *w* is nowhere preserved except in the solitary instance under Type I. All the other spellings contain *m*, which, however, may be the result of the coalescence of *w* and *n*, the *m* taking the lip-action from the former, and the nasalisation from the latter sound (cp. *Rampton* < *Rafn-*). It is, therefore, not absolutely necessary to assume the existence of an O.E. pers. n. *Bodmǣr*, of which there are no other traces, but which would correspond to the O.H.G. *Botmar* (Förstemann, 1).

It is not difficult to explain the variety of spellings recorded of this name. The *t* of Type II 3 and the modern pronunciation arose out of confusion with the noun "bottom." Intervocalic *d* seems to become open in the dialect (Type II 4 and local pronunciation). The *k* after the *s* has been assimilated, but the latter retained its voiceless quality. Various spellings show an attempt on the part of the writers to connect the second element with "hill."

BOUGHTON [būtn].

Type I.

    1086 Buchetun, D.B.

    1225 Buketon, Index.

    1316 Bucketon, F.A.

    1318 Bucton, Index.

Type II.

$\left.\begin{array}{l}1327\text{-}77 \\ 1346 \\ 1377 \\ 1535\end{array}\right\}$ Bughton $\left\{\begin{array}{l}\text{Non. Inq.} \\ \text{F.A.} \\ \text{Index.} \\ \text{Val. Eccl.}\end{array}\right.$

1571 Boughton, Index.

Type III.

1346 Button, F.A.

" The *tūn* or farmstead of *Bucca*." The phonetic development of this word presents a number of interesting features. The *k* preserved in Type I was opened before *t* (Phonology, § 20) ; the result was the first form of Type II, pronounced [buxton]. Before the *gh* [x], an *u*-glide arose which, combined with the original *u*, formed a long vowel [ū] spelt *ou* in the second form of Type II, and the modern name. Curiously enough, this *ū* does not seem to have been diphthongised, probably on account of the preceding labial (see Phonology, § 4).

Type III shows assimilation of *k* to *t* (Phonology, § 13).

BRADEBUSK (in Gonalston Parish).

c. 1500 Brodebuske, Inq. P.M. c. 1500.

" The broad bush." The second element is the Scand. *busk* (cf. *Bonbusk*). *Brode-* is a variant of *broad* ; the *a* in open syllable, found in the modern spelling, points to the influence of Scand. *breiðr*.

Of this place nothing is left but the ruins of a hospital, which derived its name "from a remarkably broad thorn tree which grew near it" (White, Directory, 1853, p. 489).

BRADMORE.

Type I.

1086 Brademere, D.B.

1216–1307 Brademar, Testa de N.

1294 Brademare, Woll. MSS.

1302 Brademere, F.A.

Type II.

c. 1500 Bradmore, Inq. P.M. c. 1500.

1534 Brademore, Index.

"The broad lake, or pool." The original *mere* of Type I was later on replaced by the more familiar *more, moor.* The first element stands for O.E. *brād,* "broad," with the vowel shortened before the combination *dm* (Phonology, § 1).

BRAMCOTE [bræmkət].

Type I.

1086 {Broncote / Brunecote} D.B.

c. 1200 Brancote, Woll. MSS.

Type II.

c. 1200 Bramcote, Woll. MSS.

1284 / 1316 } Bramcote, F.A.

Type III.

1346 / 1428 } Brauncote, F.A.

"The cot or dwelling in the place cleared of brushwood by means of fire." The first element I take to be M.E. *brand* which is recorded in the N.E.D. as meaning the "act, means or result of burning." Instances of its occurrence in pl. ns. are given by Prof. Wyld (Lancs. Pl. Ns. p. 297). On the continent this element occurs frequently in place nomenclature.

Type I retains the original *n,* the *d* having been lost between the two consonants. The *o* of the first D.B. spelling may stand for *a* before nasals (Stolze, § 2), or *u* as in the alternative spelling. This *u* is a mistake of the D.B. scribe, who was probably thinking of the rather frequent pers. n. *Brūn.* The change of *n* to *m* in Type II is due to the assimilatory influence of the initial *b* (see Phonology, § 13). Type III represents the Anglo-French pronunciation of *a* before nasals (Phonology, § 11).

N.B. It would also be admissible to derive the name from *Brand,* a pers. n. of Scandinavian origin (see Björkman).

BRECKS ("The Brecks," a tract of light forest land to the west of Boughton).

This word is of Scand. origin; it comes from *brekka,* "a slope" (Vgf.). Neither the N.E.D. nor the Dial. Dict. contains it.

BRENTSHILL (a lofty eminence covered with traces of ancient earthworks, near Barton).

"The steep hill"? With this name may be compared *Brent Knoll* near Athelney in Somersetshire which also shows traces of an old camp. The first element seems to be the dialect word (Yks., Leics.) *brent*, "steep." The *s* may be regarded as a late addition that owes its existence to the erroneous assumption that the first part of a compound pl. n. must appear in the genitive case. (Cp. *Merrils Bridge.*)

BRIDGEFORD or BRIDGFORD (East).

> 1086 Brugeford, D.B.
> 1302 Brigeford, F.A.
> 1345 Estbryggeford, Index.

"The ford by the (ruined) bridge"? The *u* of D.B. stands for O.E. *y* (Stolze, § 15). In Roman times, the place seems to have been called "ad pontem," but this is by no means established (Victoria County History, II, pp. 6, 7, 17).

It is quite true that no traces of a Roman bridge have been discovered. The original Roman bridge may have been a wooden structure which was allowed to decay in post-Roman times.

BRIDGEFORD or BRIDGFORD (West).

> 1086 Brigeforde, D.B.
> 1203 Brigiford, Index.
> 1302 Briggeford ad Pontem, F.A.

There has been a bridge at this place since the days of Edward the Elder (924), so that the ending "ford" seems somewhat out of place. It is, however, possible that "ford" means "road across a river," whatever the actual means of passing from one bank to the other may be. Or is "ford" a name that can be applied to any place on a river or brook?

BRINSLEY.

Type I.

> 1086 Brunesleia, D.B.
> 1216–1307 Brunesleg, Testa de N.
> 1291 Brunnesley, Tax. Eccl.
> 1312 Brunnesleye, Woll. MSS.

Type II.

1216–1307 Brinseley, Testa de N.
1316 Brinnesleye, F.A.

"The *lēah* or open field of *Brūn*, or *Brȳne*." The name *Brūn* is very frequent in O.E.; the variant *Brȳne* (Type II) seems to have taken its place and survived[1]. At the time of the D.B. survey, a man called *Brūn* held four bovates in this place: it is highly probable that he was the owner or settler who gave his name to the locality. In that case, Type II is due to analogy with the co-existing variant *Brȳne*. If this is correct, *Brinsley* would be one of the very few pl. ns. called after persons about whom anything is known.

BROADHOLME.

1086 Brodeholm, D.B.
1160⎫   ⎧Index.
1291⎬ Brodholm ⎨Tax. Eccl.
1428⎭   ⎩F.A.
1637 Bradham, Map in Camden.
1704 Brodham, Map of 1704.

"The broad holme, or island." The word **holm* is probably of Scand. origin. The *o* for O.E. *ā* in the D.B. entry is remarkable, as the change of *ā* to M.E. *ǭ* is not, as a rule, found as early as the date of the great survey (Stolze, § 3).

The spellings of 1637 and 1704 represent attempts at etymology. Camden's *a* in the first syllable may be due to early shortening of *ā* before *dh* (cp. *Bradmore.*)

BROUGH [braf].

"The *burh*, or fortified place." In Thoroton's History (17th cent.) the place is called *Bruff*; by that time it was "only a name." It is derived from O.E. *buruh*, a designation applied to fortified places, especially to all walled towns and camps. *Brough* is the site of the Roman station called Crocolana (MacClure, p. 109).

---

[1] It is, however, not impossible that the *u* of Type I may represent a Norman rendering of the sound of the rounded O.E. *ȳ*.

BROUGHTON (Upper) [brōtn].

Type I.

1086 Brotone, D.B.
1291 Brocton, Tax. Eccl.
1316 Brokton, F.A.

Type II.

1346 Broghton, F.A.
1571 Broughton, Index.

Type III.

1302 Brotton, F.A.

" The *tūn* or farmstead by the brook." The development of the O.E. *Brōctūn* is similar to that of *Boughton* (q.v.). Before the *kt, ō* was shortened (Type I); then *kt* became *ht* (Type II 1), then an *u*-glide developed (Type II 2). M.E. *ou* seems to be represented by ō in the modern dialect (Phonology, §9). Type III shows assimilation, *kt > tt*.

Upper Broughton occupies the eastern slope of a steep hill overlooking the Leicestershire village of *Nether* Broughton.

BROXTOW (formerly the name of a hundred; it now appears in Broxtow Hall, a farmhouse in the parish of Bilborough).

$$
1086 \left\{\begin{array}{l} \text{Brocolvestou} \\ \text{Brochelestou} \\ \text{Brolvestou} \\ \text{Bruchelestou} \end{array}\right\} \text{D.B.}
$$

c. 1175 Brocolvestou $\left.\right\}$ Woll. MSS.
c. 1190 Brogcholvestowe $\left.\right\}$

1284 Brocolstowe $\left.\right\}$ F.A.
1428 Brokestowe $\left.\right\}$

1457 Brocholwestouwa *alias* Brokestou, Index.

c. 1500 $\left\{\begin{array}{l} \text{Brokstowe} \\ \text{Brox(t)all} \end{array}\right\}$ Inq. P.M. c. 1500.

O.E. *Brōcwulfes stōw*, " the place of *Brōcwulf*." The operation of various phonetic laws has produced a vast number of more and more abbreviated forms.

Although the pers. n. involved is not recorded in the Onomasticon, it must have existed in O.E., as would appear

from this pl. n. and the one found in the Crawford Charters
(p. 70). The name is found as *Proculf* in O.H.G. (Förstemann, I).

The second element is of somewhat doubtful meaning. It is
usually employed in O.E. as signifying "place, locality"; very
often, however, its sense is that of "sacred site, burial-place"
(see Middendorf, s.v.). In the present case it may have the
latter meaning. If so, *Broxtow* would have been the burial-
place or mound of a certain *Brōcwulf*, where the men of the
hundred assembled.

Isaac Taylor explains the name as meaning "place at the
Badger's hole" which is, of course, untenable.

BUDBY.

1086 Butebi, D.B.

$\left.\begin{matrix}1278\\1316\end{matrix}\right\}$ Buteby $\left\{\begin{matrix}\text{H.R.}\\\text{F.A.}\end{matrix}\right.$

"The *bȳr* or farmstead of *Būtr*, or *Butti*." The second
element being of undoubtedly Scand. origin, the same may be
expected of the first. The names *Būtr* and *Butti* are not found
in English sources, but are recorded by Rygh (Gamle Personnavne). The change from *t* to *d* is due to the voicing influence
of the surrounding sounds.

N.B. The O.E. name *Budda* can hardly be contained in the
pl. n. seeing that the early spellings all exhibit a *t*.

BULCOTE [būkə].

1086 Bulecote, D.B.
1278 Bulkete, H.R.

1302 $\left\{\begin{matrix}\text{Bolcote}\\\text{Bulcote}\end{matrix}\right\}$ F.A.

1637 Boucot, Map in Camden.

"The cote of the bull, the cattle shed." The O.E. equivalent
of modern "bull" occurs in compounds only; this is one of
those cases. A parallel name is found in *Lambcote* (q.v.).

BULWELL [locally: bulǝl; otherwise: bulwǝl].

1086 Bulewelle, D.B.
1316 Bolewell, F.A.

The meaning of the second element is clear.   As to the first, various explanations may be offered:

(1)   It may stand for the O.E. pers. n. *Bulla*.

(2)   It may represent the O.E. **bule*, "a bull" (see prec. name).

(3)   It may describe the sound produced by the flowing water of the spring.   Although I am unable to suggest what the exact form of the O.E. name was, I feel sure that this latter is the correct explanation.   The well or spring from which the locality derives its name is still in existence and known to the people as "*the* Bulwell" without any addition; on the maps it is marked "Bulwell Spring."   There is a copious flow of brilliantly clear water rushing out of the red sandstone with a bubbling[1] noise.   In a few places the water rises from the bottom of a small pool as if it were boiling.   In the N.E.D. the noun *bulling* is recorded as occurring once, describing "the action of water issuing from a spring, bubbling."   It is there compared with French *bouillir*, Latin *bullīre*; but it is evident that both are independent onomatopoetic formations[2].   The same word is encountered in German with the addition of a frequentative *r*: *bullern*, "Blasen werfend geräuschvoll aufwallen; ein dumpfes Geräusch machen."

BUNNY.

1086	Bonei, D.B.	
1227–77⎫	Boneye	⎧Non. Inq.
1284⎭		⎩F.A.

---

[1] This word is derived from ** bullan* by reduplication, often found in onomatopoetic formations: cp. the German *surren* and the Latin *susurrare*.

[2] This particular combination of sounds is such a perfect imitation of the noise of "bubbling" water that it is often formed *ad hoc* and independently.   The following quotations will show how onomatopoetic words come into existence, affording at the same time a welcome illustration and parallel of the origin of the pl. n. under discussion.   They are taken from one of the most typical works of German Romanticism, Bettina von Arnim's "Goethes Briefwechsel mit einem Kinde" (Reclam's edition).   "...und dann die runde grüne Quelle, an der wir standen, die so ewig über sich sprudelt, bul, bul, und Du sagtest, sie rufe der Nachtigall..." (p. 287).   "Dort im Park zu Weimar gingen wir Hand in Hand unter den dichtbelaubten Bäumen, das Mondlicht fiel ein...dann führtest Du mich an die Quelle, sie kam mitten aus dem Rasen hervor, wie eine grüne krystallne Kugel, da standen wir eine Weile und hörten ihrem Getön zu.   'Sie ruft der Nachtigall,' sagtest Du, 'denn die heisst auf persisch Bulbul...'" (p. 565).

Thoroton in his History (I, p. 85) gives the correct etymology:
"Probably from Reeds." O.E. *æt bune ēȝe*, "at the water full of
reeds"; the O.E. and M.E. *bune* of uncertain derivation is
translated by "canna, harundo, calamus" in early glossaries
(see N.E.D., *bun*, sb.). The modern English meaning of *bun* is
"a hollow stem, especially of an umbelliferous plant, a kex";
compare Fitzherbert, Husbandry (1523): "The...lowe places, and
all the holowe bunnes and pypes that grow therin" (l.c.).

BURTON JOYCE.

Type I.

1086 Bertune, D.B.

Type II.

c. 1170 Birtona, Woll. MSS.

1278⎱ Birton ⎰H.R.
1291⎰   ⎱Tax. Eccl.

1302 Byrton ⎱ F.A.
1428 Birton ⎰

Type III.

1428 Burton Jorce, F.A.
1535 Burton Jorth, Joys, Val. Eccl.

Type I is one of the numerous blunders of the D.B. scribes.
It might stand for O.E. *beorg tūn*, "the farm on the hill"; but
this sense is quite different from that of the more numerous
and reliable spellings that follow.

Type II = O.E. *byrih tūn*, "the farm by the fortified place";
Type III = O.E. *buruh tūn* meaning the same thing; *buruh* and
byrih are variants of the same word.

The addition of *Joyce* is accounted for by the place having
once been in the possession of that family: "Robertus de Jorz
tenet in B." (F.A. 1302).

BURTON (West).

1086 Burtone, D.B.

1291⎫   ⎧Tax. Eccl.
1316⎬ Burton ⎨F.A.
1428⎭   ⎩F.A.

See preceding name, Type III.

BYCARDYKE.

1189 Bikeresdic, Nottm. Ch.

1278 $\left\{ \begin{array}{l} \text{Bikerisdick} \\ \text{Bikerisdik} \end{array} \right\}$ H.R.

The second element is O.E. *dīc*, "ditch," as it developed on Northumbrian territory, with *k* instead of *tf* (Bülbring, § 496). It is difficult to say what the first element is. It may stand for O.N. *bekkr*, "brook," with the nominatival *r* preserved, and the English *s* added as a sign of the genitive. It is also possible that the original compound was *bekkjar dīc*, *bekkjar* being the regular O.N. genitive form, and that later on a superfluous *s* was added when the original meaning of *er* had become obliterated. If so, the meaning would be "the dyke of the brook."

The West Riding pl. n. *Bickerton* is explained by Prof. Moorman as meaning "the enclosure by the water"; *Bickerstaffe* in Lancashire contains the same first element. Prof. Wyld translates it by "the shore of the brook" (Lancs. Pl. Ns., s.v.).

The transition from *e* to *i* in the first syllable is probably due to the influence of the following *k*: the change *bekkr*, *bekkjar* > M.E. *bicker* is similar to that from O.E. *strec*, "straight," to Northern M.E. *stric* (Morsbach, M.E. Gramm. § 109).

The modern spelling does not contain *s*, and probably goes back to the original type *bekkjar* (*bekkr*) *dīc*. It has a somewhat fantastic appearance, having been influenced by analogy of the preposition *by*, and *car*, a dialect word meaning "swamp, bog" (see *Carburton*).

CALVERTON ["vulgarly" : kōvətn ; otherwise : kāvətn, kælvətn].

1086 Calvretone, D.B.

1284 Calverton, F.A.

O.E. (Mercian) *calfra tūn*, "the enclosure of the calves." The sound development presents many interesting details. The third pronunciation recorded above is entirely based on the modern spelling.

CARBURTON or CARBERTON.

1086 Carbertone, D.B.

1278 Carberton, H.R.

"The barley-enclosure, or grange, on the marshy land, or
car." See *Barton* with which it is identical. The prefixed *car*
meaning "a pool, low-lying land apt to be flooded, boggy grass
land" (Dial. Dict.) is derived from Scand. *kiarr*, "marshy
ground."

CAR COLSTON.

> 1086 Colestone, D.B.
>
> 1216–1307 $\left\{ \begin{array}{l} \text{Kercolmston} \\ \text{Kyrkholmston} \end{array} \right\}$ Testa de N.
>
> 1284 Kercolston $\left. \begin{array}{l} \\ \end{array} \right\}$ F.A.
> 1428 Kyrkalston

"The farmstead of *Col* in the bog." The man's name *Col*
may be of O.E. or Scand. origin; Dr Björkman is inclined to
take the latter view. The Testa de N. spellings are fantastical
attempts at etymology. For *car* see preceding name.

CARLTON (near Nottingham).

> 1086 Carentune, D.B.
> 1302 Carleton, F.A.

CARLTON-IN-LINDRICK.

> 1086 $\left\{ \begin{array}{l} \text{Caretone} \\ \text{Carletone} \\ \text{Careltune} \end{array} \right\}$ D.B.
>
> 1135–54 Carletuna, Index.
> 1291 Karleton in Lyndryk, Tax. Eccl.

CARLTON (South or Little).

> 1199–1216 Karleton, Index.

CARLTON-UPON-TRENT.

> 1086 $\left\{ \begin{array}{l} \text{Carletone} \\ \text{Carentune} \end{array} \right\}$ D.B.

It is difficult, if not impossible, to say with any degree of
certainty whether the first element of these names is the Scand.
noun *carl*, corresponding to the O.E. *ceorl*, or a pers. name,
either *Carl*, or *Carla*. Dr Björkman seems to be in favour of
the latter explanation (p. 78), whereas Isaac Taylor (Engl.
Village Names, § 3), Prof. Skeat (Pl. Ns. of Beds., s.v. *Charlton*),

Prof. Wyld (Pl. Ns. of Lancs. p. 93) and Prof. Moorman (Pl. Ns. of W. Riding, pp. xvi, 42) adopt the former. According to the last named authority, the O.E. prototypes were *carla tūn, carla* being the gen. pl. of Scand. *karl*, or *carlana tūn*, with substitution of the weak gen. pl. *carlana* for the strong form *carla* (Sievers, § 257, anm. 4). The meaning of the place-names would be "the enclosure of the freemen." The meaning of *ceorl, karl* was not always that of the present-day *churl* or *carl*, which are descended from them. It was used in legal language to denote the freeman standing between the noble and the slave.

The curious early spellings noted in the above list with its omissions and transformations offer no difficulty to anyone acquainted with the vagaries of Anglo-Norman scribes.

Lindrick = "lime-wood," from O.E. *lind*, "a lime-tree," and **ric*, which seems to be identical with Low German *ricke*, "tractus, Hag, längliches Gebüsch." See Jellinghaus, who quotes the Westphalian pl. n. *Bockryck*, "beech copse."

CAUNTON.

Type I.

$$1086 \left\{ \begin{array}{l} \text{Calnestone} \\ \text{Carletun} \end{array} \right\} \text{ D.B.}$$

1166-7 Calnodeston, Pipe Roll XI.

$$\begin{array}{l} \text{c. 1200 Kalnadatun} \\ \text{c. 1216 Kalnadton} \end{array} \right\} \text{ Index.}$$

1278 Callenton, H.R.

$$\begin{array}{l} \text{1302 Calneton} \\ \text{1316 Caneton} \end{array} \right\} \text{ F.A.}$$

Type II.

c. 1225 Calfnadtun, Index.

The various forms under Type I seem to point to a pers. n. **Carlnāð* as the first element. The element *nāð* occurs in pers. ns. both in Scandinavian and West Germanic. It is, however, impossible to come to any definite conclusion unless the existence of the name **Carlnāð* could be authenticated. A reinvestigation of the pl. ns. *Kalladaberg, Kalladaland* etc., quoted by Rygh (Gamle Personnavne, pp. 155–6), whose explanation cannot be accepted, might, perhaps, throw light upon this question.

Type II seems a mere futile attempt at etymology on the part of the scribe.

CAYTHORP.

Type I.

c. 1170 Cathorp ⎤
c. 1200 Cattorp ⎦ Woll. MSS.
    1316 Cathorp, F.A.

Type II.

1216–1307 Kalthorp, Testa de N.

There is a Caythorpe in Lincs., which appears as *Carltorp* in D.B. This at once settles the etymology of the pl. n. The meaning of the prefix *carl, karl* was discussed under *Carlton* (q.v.). The development was as follows : *Carlþorp* > *Carrþorp* (Type I, which survived), or *Cal(l)þorp* (Type II); *arþ* > *āþ*, etc. (see Phonology, § 7).

CHILWELL.

Type I.

1086 ⎧ Cidwelle ⎫
     ⎪ Chidewelle ⎪
     ⎨ Cillewelle ⎬ D.B.
     ⎩ Ciluellis ⎭
1302 Chillewell, F.A.

Type II.

1284 Chelewelle, F.A.

This name also exists in Lancashire, in the modern disguise of *Childwall*, and is discussed by Prof. Wyld (Pl. Ns. of Lancs. p. 91). I take the first element to be the O.E. **celd, *cild*[1], which is not found as an independent word but appears in the Kentish pl. n. *Bapchild* (see Jellinghaus, Anglia, XX, 299; MacClure, p. 226). It is connected with Scand. *keld,* " a well, spring, pool." The second element of *Chilwell* being O.E. *wiell*,

---

[1] O. Bulg. *kladęzi,* "a well," is derived from a hypothetical Gothic noun **kaldiggs* which Dr Hirt takes to contain the root of modern Engl. *cold* (Etymologie der neuhochdeutschen Sprache, 1909, p. 45).

well, "spring," the meaning of the whole name most probably is : " The pool containing a spring, the flowing well."

The variation in the vowel of the alternative *cild* and *celd* found in Types I and II respectively is explained by assuming that the former is the Southern, the latter the Northern O.E. form (Bülbring, §§ 151, 154).

CLAREBOROUGH [klābrə].

    1086 Claureburg, D.B.
    1189 Claverburc, P.R.
    1278 Claverburg, H.R.
    1286 Clauerburge, Index.

" The fortified place where clover grows." The vowel in the O.E. *clǣfre* or *clāfre* was shortened before the combination *vr* ; the *v* was lost according to rule.

CLAYWORTH or CLAWORTH.

    1086 Clauorde, D.B.
    1155 Clawurda, P.R.
    1225 Clawrd, Bor. Rec.
    1278 Clawurh, H.R.
    1316 Clauworth, F.A.
    1637 Cloworth, Map in Camden.

" The farm in the clay land." It is remarkable that no ancestors of the first and most frequent form of this name have come down to us. All the early spellings, as well as the modern alternative, point to a shortening of the first element having taken place : O.E. *clǣȝweorþ* > *clæȝweorþ* > *claworþ* ; before the *w* an *u*-glide arose (F.A. of 1316) which formed a diphthong with the preceding vowel. This *au* had become monophthongised by the time of Camden. The principal modern form probably owes its existence to the fact that the etymology of the name had at no period become altogether obscured. The soil of this township is a rich clay.

CLIFTON (North and South) ; and CLIFTON near Nottingham.

    1086 $\left\{\begin{array}{l} \text{Cliftune} \\ \text{Clifton} \\ \text{Clistone} \\ \text{Clitone} \end{array}\right\}$ D.B.

The etymology of this name is obvious, especially to those who have visited the localities. The villages of North Clifton and Clifton near Nottingham are situated near long cliffs.

The curious *Clistone* of D.B. can puzzle only those unfamiliar with the vagaries of the Norman scribe.

CLIPSTONE.

$$1086 \begin{cases} \text{Clipestone} \\ \text{Clipestune} \end{cases} \text{D.B.}$$

1189 Clipeston, P.R.

1695 Clipstow, Map in Camden.

"The *tūn* or farmstead of *Clip.*" The man's name *Clip* is recorded once as that of a moneyer (Onomasticon).

The modern spelling seems to imply that the second element was O.E. *stān,* "stone, rock, boundary or gravestone." There is nothing in the early spellings to support this assumption ; on the contrary, the second D.B. form in particular clearly shows the second element to have been O.E. *tūn.* The modern name has simply retained the appearance given it in M.E. times by Anglo-Norman scribes who habitually rendered the Engl. *u* before *n* by *o*, often adding a superfluous *e* at the end. Camden seems to have blundered in rendering the pronunciation *Clip-stone* imperfectly.

CLIPSTONE-ON-THE-WOLDS.

See preceding name. No early forms.

CLUMBER.

1086 Clunbre, D.B.

1216–1307 Clumber, Testa de N.

White's Directory (1853) describes the appearance of the neighbourhood in the 18th century as follows (p. 586): "About a hundred years ago, it was one of the wildest tracts of Sherwood forest, being then 'little more than a black heath full of rabbits, having a narrow river running through it, with a small boggy close or two.'" Originally, the name *Clumber* belonged to a wood, from which it passed to the modern magnificent mansion and park of the Dukes of Newcastle now occupying its site.

Considering the former appearance of the locality, I take the pl. n. *Clumber* to be the same as the independent word of identical form still found in English dialects. In the Dial. Dict., the following senses of the noun *clumber* (*clumper*) are recorded : (1) "a lump, a heavy clod of earth"; (2) pl. "shapeless blocks of stone strewn over the surface of the ground"; (3) "a clump or patch of trees, plants." The N.E.D. derives this word from O.E. *clympre*, "lump, mass of metal." It is very probable that the word was originally applied (in the second sense of the Dial. Dict.) to a mass of shapeless boulders whose appearance struck the early inhabitants as sufficiently singular to characterise the site.

Modern German cognates are *Klumpen*, "unförmliche Masse," and *Klumper*, "Klümpchen."

mb is often written *nb* in D.B.; see early spellings of *Cromwell* and *Lambcote*.

COATES.

1316 Cotes, F.A.

"The dwellings, houses, or huts." The O.E. singular was *cot*, "a house, cottage."

CODDINGTON.

1086 Cotintone, D.B.
1175 Cotintona, Woll. MSS.
1316 Codington, F.A.

There exist in O.E. the pers. ns. *Cotta* and *Codda*, and it is very difficult to say which of the two is really contained in the above pl. n. The two older spellings seem to point to the former, the third to the latter. It is also possible that the original *tt* became voiced under the influence of the surrounding vowels, a process that might have been assisted by the presence of another *t* causing dissimilation.

It is equally questionable whether the *ing* is the result of the pers. n. having originally appeared in the patronymic or in the gen. sg. However, as all the old forms contain the vowel *i*, the former was most probably the case.

COLLINGHAM.

1086 Colingeham, D.B.
1189 Collingeham, P.R.
1284 Colingham, F.A.

"The home or village of the *Collings*." The pers. n. *Coll*, *Coll(l)a* etc. is comparatively frequent in late O.E. records. Dr Björkman is of opinion that it came from Scandinavia.

The same patronymic seems to occur in the continental pl. n., O.H.G. *Collinchova* (Förstemann, II).

COLSTON-BASSET [kousn, or less frequently koulsn].

1086 Coleton, D.B.
1160 Colestun, Index.
1302 Colston Basset, F.A.

"The *tūn* or farm of *Col*." See *Car Colston*. *Basset* is the name of a noble family that once held land in this place; see *Bassetlaw*.

COLWICK [kolik].

1086 $\begin{cases} \text{Colewic} \\ \text{Colewi} \\ \text{Colui} \end{cases}$ D.B.

1225 Colewic, Bor. Rec.

The first element of this name is undoubtedly the pers. n. *Col*, which also occurs in *Colston* (q.v.). What the second part means, it is difficult to say for certain. If the first element is of Scand. origin, as Dr Björkman assumes, the second may be expected to be derived from the same source. In that case, -*wick* would go back to the O.N. *vīk*, "creek, bay," and "*Col's* creek" would be the interpretation of the modern pl. n.

There also exists, in the English language, the word *wick*, "farmstead village." It goes back to O.E. *wīc*, "dwelling-place, village," but is found in that form in the Northumbrian dialects only, the Southern and Midland type being *wīch* (in pl. ns. only), with the *c* (*k*) fronted. (See Bülbring, § 496.) Cp. *Papplewick*. If the O.E. *wīc* occurred in this county it would have the latter form unless it could be proved that it was imported from north of the Humber.

COSSAL.

1086 Coteshale, D.B.
c. 1200 Cozale, Woll. MSS.
1284 Gossale ⎫ F.A.
1302 Cossale ⎭

"The *healh* or valley of *Cot*(*ta*)." The *z* in the Woll. MSS. spelling stands for *ts* according to Norman-French practice. *ts* has become *ss* through assimilation. The change from initial *c* to *g* in the first F.A. form is by no means an isolated one; cp. the spelling of *Cotgrave* in D.B.

COSTOCK or CORTLINGSTOCK.

1086 ⎧ Cortingestoche ⎫ D.B.
      ⎩ Cotingestoche ⎭
1166–7 Cordingestoch, P.R.
1302 Cortelingstocke, F.A.
1535 Cortelyngstoke, Valor Eccl.
1637 Corthigstoke, Map in Camden.

"The dwelling-place or village of the *Cortlings*," O.E. *Cortlinga stoc*. For the exact meaning of *stoc* see List of Elements. The pers. n. *Cortel* is not recorded in any of the collections of names. It must, however, have existed. The O.E. *Cyrtel* is found in the Crawford Charters (p. 52), and the editors have added further instances of pl. ns. containing it. See also Skeat, Pl. Ns. of Cambridgeshire, s.v. *Kirtling*. I take both names to be derived from an older **Curt* (cp. *Crotus*, *Crotilo*, Werle, Index), by means of the diminutive suffix *il, al*; the addition of *il* produced *Cyrtel*, whereas *Cortal* gave rise to *Cortel*. On the continent, the same pers. n. seems to be contained in the Low German *Krotillandorf* (Förstemann, II). The varying position of *r* is easily explained, as metathesis in pers. ns. is not infrequent.

The name *Curt* may be identical with the Germ. adjective *kurz*. The surname *Körtzel* is found in modern German.

COTGRAVE.

Type I.

1086 Godegrave, D.B.
1157 Cottegaua.
1230 ⎫ Cotegrave ⎧ Bodl. Ch. and R.
1284 ⎭   ⎩ F.A.

Type II.

? Cotesgrafe {Reg. Lenton Abbey, quoted
by Thoroton, I. 166.

"At the grave of *Cot(ta)*, O.E. *æt Cottan* (Type I) or *Cottes*
(Type II) *græfe.*"

COTHAM.

$$1086 \left\{ \begin{array}{l} \text{Cotun} \\ \text{Cotes} \end{array} \right\} \text{D.B.}$$

1302 Cotum, F.A.

"At the dwelling-houses or cottages." The D.B. spellings
represent the O.E. dat. pl. and nom. pl. respectively. The
former survived (from O.E. *æt cotum*), the final *m* being later
taken to stand for *-ham, -home*, as usual. Cp. *Coates, Cottam.*

COTTAM (under Leverton).

1302 Cotum, F.A.

The same as *Cotham*, without the erroneous etymological
spelling, as far as the additional *h* is concerned.

CROMWELL [krɑməl, but usually kromwəl].

1086 Crunwelle, D.B.
1278 ⎫   ⎧ H.R.
1302 ⎬ Crumwell ⎨ F.A.
1637 ⎭   ⎩ Map in Camden.

"At the winding brook," O.E. *æt crumb(um) welle*. From O.E.
crumb, "winding, crooked," and *well*, "a brook." The development of sounds is well in accord with the general rules. The
original, and natural pronunciation of this pl. n. is hardly known
outside the immediate neighbourhood. In Ireland, however, the
Protector is still called [krɑməl], which seems to prove that
hatred has a better memory than love or admiration. Or does
it merely show that the Irish have not yet come under the
spell of the printed word to the same extent as their English
brethren?

CROPWELL BISHOP, and CROPWELL BUTLER.

1086 $\begin{cases} \text{Crophille} \\ \text{Crophelle} \end{cases}$ D.B.

1216–72 Cropil, Index.

(a) 1316 Croppehull Episcopi, F.A.
    1336 Crophull Bisshop, Index.

(b) 1284 Crophill Botiller $\Big\}$ F.A.
    1302 Croppilboteler

    1368 Crophull Botiler, Index.

c. 1500 Cropwell, Inq. P.M. c. 1500.

"At the hump-shaped hill." The first element is the Scand. *kroppr*, "a hump or bunch" (Vigf.); it occurs in pl. ns. found in the Landnama Bok. The modern spelling is due to confusion with *well*, which undoubtedly was assisted by the development of a labial glide after the *p* when the *h* had been lost. The additions explain themselves: the Archbishop of York and the noble family of Butler were at one time the respective owners of the two villages.

Thoroton gives the correct etymology of this name when he says (I, 189) that they (viz. C.-Bishop and C.-Butler) were so named "from a round Hill which is between them, now called *Hou-Hill*."

CUCKNEY.

Type I.

1086 Cuchenai, D.B.
1200 Cucheneia, Bodl. Ch. and R.
1278 Cuckenay, H.R.
1302 Cockeney, F.A.
1329 Kukeney, Index.

Type II.

1250 (Richard de) Kukeney (in)
     Kuyekeney, Bodl. Ch. and R.

"At the quick, running water, or brook." There must have been two O.E. alternative forms of this name: *æt cucan ēʒe* (Type I) and *æt cwican ēʒe* (Type II), *cucu* being a variant of *cwicu,* "quick, alive." (See Bülbring, § 464.) With this name we may compare the Low German *Quickborn,* and the O.H.G. *kecprunno,* "lebendiges Wasser, Quelle," in the poem of Christ and the Samaritan Woman (14: "uuâr maht thû guot man, neman quecprunnan?" The Vulgate has: "unde ergo habes aquam vivam?" St John iv. 11).

The second element is O.E. *ēȝe,* "water, river, stream."

DALBY (brook near Hickling).

DALINGTON [dælintn].

     1086 Dallingtune, D.B.

     1155 ⎫
     1156 ⎬ Derlintun, P.R.
     1157 ⎭

"The *tūn* or farmstead of the Dēorlings." The D.B. form shows assimilation of *rl* to *ll* which is quite in accordance with the almost universal practice of its compilers. On the other hand, it is surprising to find *a(r)* for *er* so early as the date of D.B. The phonetic development of *erl* is treated in the Phonology (§ 7).

DANETHORPE.

     1086 Dordentorp, D.B.

     1637 Dernthorp, Map in Camden.

     ? ⎧ Dernethorpe ⎫ quoted by Thoroton from
       ⎨ Dornethorp ⎬ unknown sources.
       ⎩ Darnethorp ⎭

"The *thorp* of *Deorna.*" In D.B., O.E. *eo* is represented by *o*; the pronunciation probably was [œ], mid-front-wide-round. The O.E. prototype was *Deornan þorp,* of which the first *n* had become denasalised in the pronunciation of the Normans under the dissimilatory influence of the following *n,* the result being the voiced point stop, *d.* Zachrisson (pp. 120 sqq.) has collected a number of similar changes.

The development of *er* (> *ar,* > *ā* > *ei*) is treated elsewhere (Phonology, § 7).

DARLTON [dāltn].

1086 Derluveton, D.B.

$\left.\begin{array}{l}1278 \\ 1316\end{array}\right\}$ Derleton $\left\{\begin{array}{l}\text{H.R.} \\ \text{F.A.}\end{array}\right.$

1695 Darleton, Map in Camden.

"The *tūn* or farmstead of *Dēorlāf*." The D.B. spelling is the most reliable in this case. There existed another *Dēorlāfestūn* in O.E. (Cod. Dipl. 1298), which has resulted in the modern *Darliston*, Staffs. The existence side by side of these two names—one with and the other without *s*—clearly demonstrates that the genitival *s* may be absent even if the first element is a pers. n. following the strong declension.

DAYBROOK (under Arnold).

Apparently a modern name derived from that of the brook on which the hamlet is situated. There used to be cotton mills in this neighbourhood worked by water-power. It is said that the brook was frequently stopped during the night, so that the water might accumulate for the day's work: thus the brook carried water in the daytime only. I give this somewhat singular account as it was related to me. This explanation is very doubtful.

DEAN (brook near Hickling).

This name may be of Celtic origin; there is a river *Dane* in Staffordshire, and another in Cheshire. Isaac Taylor (Words and Places, ch. IX) enumerates a large number of river-names containing a similar element found in various parts of Europe formerly occupied by Celts.

DEVON (river) [dīvn].

1680 Devon, Index.

A Celtic river-name; see *Dean*.

DOVER BECK.

1225 Doverbec, Bor. Rec.

This name is a tautology. The first part is of Celtic origin, having a common ancestor with modern Welsh *dwr*, *dwfr*, "water." When the original meaning was lost, an explanatory superfluous *beck*, from Scandinavian *bekk(r)*, "brook," was added.

DRAKEHOLES.

This small hamlet, of which no early forms are available, is situated "in a narrow part of the hills through which the Chesterfield canal passes by means of a tunnel." The meaning may be: "the holes of the dragon, or dragons," O.E. *dracan*, or *dracena hol*(*as*). Cp. Low German *Drakenloch*, a small valley leading out of the Urpetal (Jellinghaus, p. vi).

DRAYTON (East and West).

$$1086 \left\{ \begin{array}{l} \text{Draitone} \\ \text{Draitun} \end{array} \right\} \text{D.B.}$$

$$1316 \left\{ \begin{array}{l} \text{Est} \\ \text{West} \end{array} \right\} \text{Drayton, F.A.}$$

"The hidden *tūn* or farmstead." There exist numerous Draytons all over the country. Prof. Skeat (Pl. Ns. Cambs. p. 9) was the first to explain their etymology. He compares the element *drai* with a squirrel's *dray*, which he says is derived from an O.E. *dræʒ* with the wider sense of "retreat, hiding-place."

The appearance of the modern village of West Drayton seems to bear out the suggested etymology. West Drayton is situated in low-lying, boggy country, surrounded on all sides by distant hills. It is not discovered by the traveller until he has come close upon it. When dwellings were low and brushwood and trees more plentiful, the seclusion of the spot must have been still more apparent. East Drayton, however, is quite open.

DRINSEY NOOK.

The "Nook" is a slight projection of dry land among meadows subject to floods. It was an island before that part of the country was drained. In the absence of early spellings one can only guess at the meaning. I take it to have been O.E. *Drenges ēʒe*, "Dreng's island." The word *Dreng* is used both as a pers. n., and an appellative. The latter, which is the original of the pers. n., comes from Scandinavian and designates members of that class sometimes spoken of as *rādcnihtas* in O.E. literature (Pl. Ns. of the W. Rid. p. xxiii). They were the successors of the old *þeʒnas*, "retainers of a chief, noblemen," principally employed in warfare. Above the *ceorl* (see *Carlton*) in station they often enjoyed the privileges of the mediæval nobility (see Vinogradoff, Growth of the Manor, p. 220; Engl. Society in the 11th Cent. p. 62).

It is doubtful whether the word is used in the first or second sense here. The West Riding pl. n. *Dringhouses*, older *Drengehous*, most probably contains the appellative in the plural : *drenga hūs*, "the houses of the soldiers or noblemen."

ey > iy according to rule; the form *Dring* is found in M.E. (Phonology, §6). *ys > ns* through assimilation (Phonology, §13).

DUNHAM-ON-TRENT.

1086 Duneham, D.B.
1155 Dunehā, P.R.
1316 Dunham, F.A.

The second element of this name is undoubtedly O.E. *hām*, "home, homestead, village." The first may be either O.E. *dūn*, "hill, mountain," or the genitive of the pers. n. *Dun(a)*. The place being situated on a gentle eminence, the former alternative may be taken as the most likely interpretation. Thus the meaning would be "the village on the hill."

It has been suggested that this place derives its name from the "dunes," i.e. hills of blown sand found in the neighbourhood. This is impossible, as the word *dune* meaning "a low sand-hill" is of comparatively recent introduction, having come into the English language through the medium of French speech. The N.E.D. gives 1213 as the first date of its occurrence.

DUNSELL-IN-TEVERSAL.

There are no early forms. It may be derived from O.E. *dūn seld* or *setl* (Bülbring, § 444), "the dwelling on the down or hill." But this is a mere guess for which there is no reliable evidence. However, the farmhouse bearing this name is actually situated on a hill.

EAKRING.

Type I.

1086 { Eceringhe / Echeringhe } D.B.
1174–5 Ekeringa, P.R.

Type II.

1637 Akringe, Map in Camden.

1704 Akring, Map 1704.

Type III.

1156 Eikeringe ⎱ Index.
1200 Eigring ⎰

1278 Aykering, H.R.

1291 { Eycring ⎱ Tax. Eccles.
      { Eykringk ⎰

1302 Eykring ⎱ F.A.
1428 Aykering ⎰

In spite of the numerous early spellings it is impossible to explain the etymology of this name with any degree of certainty. I take it to have been a patronymic family name in the gen. pl. followed by some such word as *hām* or *tūn*. It may have been *Eādwæceringa hām*, "the home of the descendants of *Eādwæcer*." This very long name would most certainly be shortened and its pronunciation might have been simplified at a very early period, through the following stages : *ēadceringa* > *ēaceringe* (> modern Eakring), or *æcringe* (> Type II), or with later shortening of the initial vowel, *ecringe* (> Type I).

The forms quoted as Type III I take to be due to popular etymology. The pl. n. was explained as *eikar ing*, "the meadow of the oaks," and I am told that this interpretation fits the locality very well. *eikar* is the Scand. gen. pl. of *eik*, "oak tree," *ing* the M.E. form of *eng*, "meadow," derived from the same language (see Pl. Ns. of the West Rid. p. xl).

EASTWOOD.

Type I.

1166–7 Est Twait, P.R.

c. 1200 { Estweit ⎱ Woll. MSS.
        { Hestweit ⎰

1225 Estwaite, Bor. Rec.

1227–77 ⎱ Estweyt { Non. Inq.
1316   ⎰   { F.A.

1483 Estwyt
1495 Estwhaite ⎱ Index.
1590 ⎱ Eastwayte ⎰
1642 ⎰

Type II.

? (Eswaicte or) Eastwood, Thoroton, II, 236.

Type III.

1086 Estewic, D.B.

"The *thwaite* or outlying farm in the East." The second element is of Scandinavian origin : O.N. *þveit*, "piece of land, paddock, parcel of land ; originally used of an outlying cottage with its paddock." As it is of very rare occurrence in this county, it was replaced by the more familiar *wood* (Type II).

The D.B. form is due to a misreading of *c* for *t*, these two letters being frequently interchanged on account of graphic resemblance.

EATON.

Type I.

1086 $\begin{cases} \text{Etune} \\ \text{Etone} \\ \text{Ettone} \\ \text{Ættune} \end{cases}$ D.B.

1302 Eton, F.A.

Type II.

(Eaton or) Idleton, White, Directory, 1853.

"The *tūn* or farmstead by the river (Idle)." The O.E. proto-type seems to have been : *īdel ēa tūn*, "the farm by the brilliant river" (see *Idle*, p. 72). The distinctive addition to the river could be left out, as it was certainly known in the neighbour-hood as *sēo ēa*, "the river," pure and simple.

EDINGLEY [ediŋli].

1302 Edingley, F.A.
1637 Heddingley, Map in Camden.

"The *lēah* or field of the sons of *Eda*, or of *Eada*, or of *Eādwin*." Both O.E. *Edinga lēah* and *Edanlēah*, as well as *Eādwin(es) lēah*, might result in the modern form.

Camden's spelling proves that the *e* was pronounced short; the initial *h* means nothing.

EDWALTON [edw'ōltn, edltn].

1086 Edwoltun, D.B.
1302 Edwalton, F.A.

" The *tūn* or farm of *Eādweald.*" The absence of the genitival *s* is noteworthy. Of the two pronunciations recorded, the second is the only natural one. The shortening of the initial vowel (before *dw*) and the subsequent loss of *w* beginning an unstressed syllable are in accordance with well established rules.

EDWINSTOWE or Edwinstow or Edenstowe (Edensta, Hope).

Type I.

1086 Edenestou, D.B.
1278 Edenstow, H.R.
1428 Ednestowe, F.A.
c. 1500 } Edenstow { Inq. P.M. c. 1500.
1637 } { Map in Camden.

Type II.

c. 1500 Eddingstow, Inq. P.M. c. 1500.

It is very strange that the old spellings do not contain a *w* in a single instance. The etymology seems to be : " The *stōw* or place of *Eādwine.*" It is said that King Edwin's body was brought to Edwinstowe after the battle of Heathfield, A.D. 633, " and from what we know of this obscure period it does not seem unlikely that such may have been the case" (Guilford, p. 84). This view is confirmed by the meaning of the suffix *stōw*, " a holy place, sanctuary, sepulchre." In the present name it may have designated the burial-place of the king. See Middendorf on *stōw*, and cp. *Broxtow.*

Type II shows an interesting development of unstressed (*w*)*in*, on which see Alexander, The Suffix *ing.*

EGMANTON [egmɘntn].

1086 Agemuntone, D.B.
1278 } Egmanton { H.R.
1302 } { F.A.

" The *tūn* or farm of *Eċgmund.*" The O.E. *ġ* should result in modern *dg* (*Edgmanton). The present form is either a mere spelling, or the fronted $\dot{g}$ has been changed to the back-stop under Scandinavian influence.

The development of the unstressed u is interesting: $u > o > a > \partial$.

ELKESLEY [locally : el(k)sli ; otherwise : elkəsli].

1086 $\left\{ \begin{array}{l} \text{Elchesleig} \\ \text{Elchesleie} \\ \text{Elcheslie} \end{array} \right\}$ D.B.

1278 Elkesle, H.R.

1316 Elkesley, F.A.

c. 1500 $\left\{ \begin{array}{l} \text{Ellersley} \\ \text{Elkesley} \\ \text{Elsley} \end{array} \right\}$ Inq. P.M. c. 1500.

$\left. \begin{array}{l} 1599 \\ 1637 \\ 1704 \end{array} \right\}$ Elsley $\left\{ \begin{array}{l} \text{Map.} \\ \text{Map in Camden.} \\ \text{Map.} \end{array} \right.$

Probably : " the *lēah* or field of *Ealc*." This pers. n. is not recorded in the Onomasticon ; we find, however, *Ealac*. It must be the abbreviated form of a full name beginning with *Ealh-*, *Ealc-*. If the first spelling taken from the Inq. P.M. is not merely an erroneous one, the original pers. n. involved would have been either *Ealhheard* or *Ealhhere*, both of which are remarkably frequent in O.E. That a very primitive form of a pl. n. should be employed by a writer about the year 1500 is not so improbable as it would seem at first sight ; for he may have been drawing from local documents of great antiquity, which are now lost.

On the name *Ealc* see Staffs. Pl. Ns., s.v. *Elkstone*.

ELSTON.

Type I.

1086 Elvestune, D.B.

Type II.

1302 Eyliston, F.A.

It is difficult to give the etymology of this name on account of the scarcity of early spellings. The most probable derivation seems to be from *Eilafes tūn*, " the farmstead of *Eilaf*." This assumption would at once explain the *v* in Type I and the *ey* in Type II. The pers. n. involved is Scandinavian in origin. (See Björkman, s.v.)

ELTON.

1086 Ailtone, D.B.
1284 Elton, F.A.

"The *tūn* or farm of *Æȝel.*" This pers. n. is a late, probably Norman variant of the older *Æðel.* The relation between the two has been fully discussed by Dr Zachrisson (pp. 100 sqq.).

EPPERSTON.

$$1086 \left\{ \begin{array}{l} \text{Eprestone} \\ \text{Epstone} \end{array} \right\} \text{D.B.}$$

$$\left. \begin{array}{l} 1225 \\ 1302 \end{array} \right\} \text{Epreston} \left\{ \begin{array}{l} \text{Bor. Rec.} \\ \text{F.A.} \end{array} \right.$$

"The *tūn* or farm of **Eopeorht.*" This pers. n. is not recorded in the Onomasticon. It may, however, be safely assumed to have existed as the corresponding continental form *Eoberht* exists (Förstemann, 1). It is derived from *eoh-berht*, which explains the *p* in the English type : this arose out of the *b* through the unvoicing influence of the preceding *h* which disappeared. Cp. Scotch *neeper*, "neighbour," from O.E. *nēahgebūr*, which exhibits a similar change.

EREWASH (river).

c. 1175 Yrewis, Woll. MSS.
$$1637 \left\{ \begin{array}{l} \text{Erewash, Camden, p. 550.} \\ \text{Arwash, Map in Camden.} \end{array} \right.$$

The termination is identical with the modern dialect word *wash*, "any shore or piece of land covered at times by water; a mere ; an inundation" (Dial. Dict.), which is also the name of the well-known arm of the sea between Norfolk and Lincoln, the Wash. The word is, no doubt, connected with the root contained in *to wash* and *water*.

The first element may be a pre-English river-name **Ir-*, **Er-*, which seems to be contained in the Lancashire *Irwell.* But this is very doubtful.

EVERTON.

> 1086 Evretone, D.B.
> 1189 Everton, P.R.

"The *tūn* or farmstead of *Eofer*." Note the absence of the genitival *s*. The O.E. pers. n. *Eofer* and its compounds are comparatively frequent.

FARNDON.

Type I.

> 1316 Farnedon ⎫
> 1402 Ferndon ⎬ F.A.

Type II.

> 1086 Farendune, D.B.
> 1392 Farendon ⎫
> 1543 Farunden ⎬ Index.
> 1586 Faringdon ⎭

Type III.

> 1637 Farmdon, Map in Camden.

In spite of Type II, I take this name to be derived from O.E. *fearn dūn*, "fern-covered hill." Type I represents the original form without any unusual alteration, *er* being merely a spelling of *ær* into which both M.E. *ar* and *er* had developed. As to Type II, I assume that a vowel-glide arose between the *r* and *n* which is represented by various symbols. *Farendun* then developed into *Faringdon* (1586), a change which has been discussed on another page (Phonology, § 13).

Camden's spelling is either a blunder or shows that the *n* had become assimilated to the initial *f*. See Phonology, § 13.

FARNSFIELD.

Type I.

> 1189 Farnefeld, P.R.
> 1637 Farnfelde, Map in Camden.

Type II.

> 1086 ⎰Farnesfeld⎱ D.B.
>   ⎱Franesfeld⎰
> 1331 Farnesfeld, Index.

M.

4

Type II seems to imply a pers. n. (O.E. *Fræna*, with metathesis?) as the first element on account of the genitival *s*. However, I give preference to Type I, explaining the name as O.E. *fearnfeld*, "the open space, or plain covered with fern." The *s* was introduced on the mistaken notion, derived from analogy, that the first element was a pers. n. Dr Zachrisson gives numerous instances of a similar "loose" or inorganic *s* (pp. 118 sqq.).

FELLEY.

1240 Felley, Bor. Rec.

This may stand for O.E. *feld lēah*, dat. *feld lēaʒe*, "the open field," or "the field in the plain." The place is situated partly on a lofty eminence, and it seems probable that the name originally referred to the lower part of the locality. The disappearance of *d* between the *l*'s is natural.

The view expressed above might be confirmed by the fact that there existed an exactly similar name in Friesland— *Veldlagi*, quoted by Förstemann (II).

FENTON.

1086 $\begin{cases} \text{Fentune} \\ \text{Fentone} \end{cases}$ D.B.
1316 Fenton, F.A.

Considering the geographical position of this place, there can be no doubt as to its etymology. "The farm in the fen," from O.E. *fenn*, "mud, dirt, fen." The country is drained now.

FINNINGLEY (Finlah, Hope).

Type I.

1086 Feniglei, D.B.
1428 Fenyngley, F.A.

Type II.

1278 Finningelay, H.R.
1302 Finningley $\Big\}$ F.A.
1316 Fynyngeleye

Two explanations of this name are possible according to which type is considered original.

(1) "The *lēah* or field of the *Finnings*, the sons of *Finn.*" This pers. n. is of Scandinavian origin, and always borne by Norsemen, either in history or fiction. A family of the same patronymic is found in Southern Germany: *Finninga* quoted by Förstemann (II).

(2) I am myself inclined to believe that Type I represents the original name more truly than Type II which is derived from the former. O.E. *fenninga tūn*, "the farm of the dwellers in the fen," is a most appropriate name for the locality situated in the old marshes. The suffix *-ing* was used in Germanic to derive the name of a tribe from the locality they inhabited. The best known examples of this usage are the designations of the two nations into which the Goths were divided[1]. The Ostrogothi were called—in Latin garb—*Greutingi*, because they inhabited sandy plains (cp. O.E. *grēot*, "sand"), whereas the Visigothi were known as *Tervingi*, because they lived in a country covered with woods (cp. O.E. *trēow*, "tree"). An O.E. example of a corresponding derivation (of which there are many more) is quoted from a charter by Dr Middendorf, s.v. *hēah*: *hēantunninga gemǣre*.

The transition from *e* to *i* before *n* is not without parallel in the English dialects. Instances of such a change are quoted in the Dial. Grammar, § 55. See also Horn, § 38.

FISKERTON.

Type I.

1086 Fiscartune, D.B.
1278 Fiskerton, H.R.
1316 Fyskerton, F.A.

Type II.

1278 Fiskiston, H.R.

"The *tūn* or farm-house of the fisherman, or fishermen." The O.E. original was either *fiscera tūn* (gen. pl., Type I), or *fisceres tūn* (gen. sg., Type II). In the ordinary course, the O.E. *sc* should be represented by *sh* in the M.E. and modern forms.

---

[1] Zeuss, Die Deutschen und ihre Nachbarstämme. München, 1837.

The *sk* is explained by assuming influence of Scand. *fiskr*, "fish," which may also have been used as a pers. n.

To this very day this village is a favourite resort for fishermen.

FLAWBOROUGH.

Type I.

1086 Flodberge, D.B.

Type II.

1316 Flaubergh, F.A.

The second element is the O.E. *beorg*, "mountain, hill, mount," the modern spelling of which has been influenced by the more frequent *borough*, the representative of O.E. *burh*. This part of the name is explained by the fact that the little village is situated on a hill.

I can make nothing of the first element, which appears in two strange spellings.

FLAWFORD (under Ruddington).

There are no early forms. The first element seems to be identical with that contained in the preceding pl. n.

FLEDBOROUGH.

1086 Fladeburg, D.B.

1278 Fleburg, H.R.

1302 Fledburgh, F.A.

"The *burh* or fortified place of *Flǣda*." An O.E. *Flǣdanburg*, which might have been the ancestor of the modern name, is recorded in a charter (Cart. Sax. 76,238).

FLEET (river).

From O.E. *flēot*, "stream, channel"; the word is connected with *flēotan*, "to float, sail, swim." In modern H.G. we find its cognate *Fliess*, "small river," M.H.G. *vliez*. The Low German form is *Fleet*, from older *vlēt*.

FLINTHAM.

1086 {Flintham / Flinteham} D.B.

1284 Flintham, F.A.

From O.E. *flint*, "rock," and *hām*, "home, habitation, village." It is difficult to say what is the exact sense intended to be conveyed by the composition of these two elements. It might have been the "house in the rock, or by the rock, or built of rock" etc.

FULWOOD.

Although there are no early forms, the etymology of this name is quite clear. The O.E. ancestor is *fūl wudu*, "the foul, dirty, boggy wood." Before the combination *lw*, the *ū* was shortened, whereas in the independent word, *foul*, it remained long and was diphthongised.

GAMSTON (near West Bridgford).

> 1086 Gamelestune, D.B.
> 1302 Gameleston, F.A.

GAMSTON (near East Retford).

> 1086 Gamelestune, D.B.
> 1278 Gameleston, H.R.

"The *tūn* or farmstead of *Gamal*." The D.B. forms preserve the O.E. appearance of this name almost completely. The pers. n. is of Scandinavian origin.

GATEFORD.

Type I.

> 1278 Gaytford, H.R.
> c. 1500 Gaytforth, Inq. P.M. c. 1500.

Type II.

> c. 1500 Gatford, Inq. P.M. c. 1500.

"The goat ford." From O.E. *gātaford*, which, with the *ā* shortened at an early period before the combination *tf*, resulted in Type II. The other type preserved in the modern spelling exhibits the influence of Scand. *geit*, "goat." See Björkman, Scand. Loanwords, p. 42. The *th* instead of *d* also points to Scandinavian origin; cp. O.N. *fjorðr*.

Fords are often named after the animals that passed through them, as *Oxford*, *Swinford*, *Hertford*.

There is a *Gateforth* in the West Riding wholly Scandinavian in appearance. Dr Moorman interprets it as "the ford of the goats." The same name is borne by a place in O.H. German territory, *Geizefurt* in Hesse (Förstemann, II).

GEDLING.

Type I.

1086 Ghellinge, D.B.
1278 Gedling, Kedling, H.R.
1302 Gedling ⎱ F.A.
1346 Godeling ⎰ F.A.

Type II.

1189 Gedlinges, P.R.

Type III.

1637 Gadling, Map in Camden.

I take this name to be derived from O.E. *on gædelingum*, "among the companions in arms" (Type I). The nominative plural *gædlingas* is the ancestor of Type II. Camden's *a* (Type III) goes back to O.E. *æ*, whereas the other forms contain *e* which most probably is due to a kind of secondary *i*-mutation (see *gædeling* in Sievers and Bülbring, Indices). In M.E. both *gedeling* and *gadeling* are found (N.E.D.).

Gilling in Yorks. apparently has the same origin; it occurs as *in* (*on*) *Getlingum, in Gætlingum* in Bede (Miller, O.E. Bede, p. 43). On the continent are found O. Low Germ. *Getilingthorp* and O.H.G. *Gellingin*, modern Göllingen near Sondershausen (Förstemann, II).

GIBSMERE.

1086 Gipesmare, D.B.
1302 Gyppesmere, F.A.

The second element is O.E. *mere*, "lake, pool." In the D.B. form, the *a* is due to the influence of Scand. *mar*, "the sea." This substitution of *mare* for *mere* in D.B. is particularly noticeable in Yorks. pl. ns. (See Stolze, § 7, anm. 3; Jellinghaus, p. 307.)

The first element seems to be a pers. n. It may be identical with the one contained in O.E. *Gypeswīc*, modern *Ipswich* (Cod.

Dipl., Index). The initial *g* of the O.E. name must have been fronted, so that the *y* cannot but represent an original West Germanic palatal vowel; otherwise the disappearance of the initial *g* in *Ipswich* could not be explained. The unfronting of this sound in the Notts. pl. n. seems to be due to Scandinavian influence which has also been at work in the second element. The meaning is "Gippes pool." The pers. n. involved is not recorded in the Onomasticon. It may be contained in the Norwegian pl. n. *Gipsen* discussed by Rygh (N. Gaardnavne, p. 345).

The combination *ps* was changed to *bz* under the influence of the surrounding voiced sounds.

GILTBROOK (under Greasley).

In the absence of early spellings nothing definite can be said about this name. It may stand for O.E. *gylden brōc*, "the golden, i.e. yellow brook." The first element may have been influenced by the adjective *gilt*. Similar river-names occur on the continent; O.H.G. *Goldaha*, O. Low Germ. *Goldbiki*, O.H.G. *Goldgieze* are enumerated by Förstemann (II).

GIRTON.

1086 Gretone, D.B.
1278 ⎫   ⎧ H.R.
1291 ⎬ Gretton ⎨ Tax. Eccl.
1316 ⎭   ⎩ F.A.
1704 Girton, Map.

"The *tūn* or farmstead in the sand," O.E. *grēot tūn*. The country round Girton is very marshy and sandy in places, and both to the north and south of the village considerable dunes have been formed by the drift sand from the Trent. O.E. *grēot* means "sand, rubble," and is identical with O.H.G. *grioz*, modern Germ. *Gries*. The Hessian pl. n. *Griesheim* has the same meaning as Girton.

The phonetic development is quite regular: M.E. *ē* < *ēo* is shortened before *tt*; metathesis of *r* is characteristic of the dialect (Phonology, § 15); the modern spelling *ir* represents the sound into which *er* had developed.

GLAPTON.

1216–72 Glapton, Cal. Inq. P.M. I.

Probably from O.E. *Glæppan tūn*, "the farmstead of *Glæppa*."
The pers. n. is found in O.E., and also in O.H.G. as *Claffo*,
Clapho etc. (Förstemann, I); or possibly the Scand. name *Clapa*,
traced as occurring in England by Dr Björkman, may be contained in this pl. n.

GLEADTHORPE GRANGE (under Worksop) [gledþōp].

1086 Gletorp, D.B.
1278 Gledetorp, H.R.
1853 Gledthorpe, White's Directory, p. vi.

It has been suggested that the first element is O.E. *glǣd*,
"bright, clear, glad"; the second is Scand. *þorp*, "village,
hamlet." The meaning would be "the bright, pleasant hamlet."

It is, however, very doubtful whether the variant **glǣd*,
containing a long vowel, ever existed. It is said to occur in
poetic texts, and might be represented in M.E. by the spelling
glead (N.E.D., s.v. *glad*). I prefer to leave the etymology of the
first element doubtful, and would refer investigators to similar
names: Gleadless, Gledhow, and Gledstow Hall, all in the
West Riding of Yorkshire.

GOLDTHORPE (under Hodsock).

There are no early spellings. I take this to be derived from
Goldan þorp, "the hamlet of *Golda* or *Golde*." Whether this pers.
n. is of O.E. or Scand. origin I am unable to say.

GONALSTON [gɑnəsn].

(1)	1086	Gunnulfestone	D.B.
(2)		Gunnulvestune	
(3)	1278	Guneliston	H.R.
(4)		Gunohston	

(5) 1302 Gunnolston ⎫
(6) 1316 Gonelston ⎬ F.A.
(7) 1346 Gonaldeston ⎭
(8) 1637 Gumalston, Map in Camden.

O.E. *Gunnulfes tūn*, " the farmstead of *Gunnulf.*" The latter is a Scandinavian man's name. No. 1 evidently is copied from an O.E. document with *f* instead of the M.E. *v.* The *h* in No. 4 and the *m* in No. 8 are scribal blunders. The *d* in No. 7 is due to confusion with the well-known Scand. female name *Gunnhild.*

The history of the unstressed vowel in the second syllable is interesting : *u > o > a > ǝ.*

GOTHAM [goutm].

1086 Gatham, D.B.
1152 Gataham, Index.
1316 Gotham, F.A.

O.E. *gāta hām*, " the home of the goats, the goat village." The Index form is evidently copied from an O.E. document. Thoroton (1, 36) has the following explanation : " a Dwelling or Home of Goats."

GOVERTON (under Bleasby) [gouvǝtn].

1302 Goverton, F.A.

O.E. *Gārfriðes tūn*, " the farmstead of *Gārfrið.*" This pers. n. is found once in an O.E. document. It is very frequent on the continent, as *Gairfrid* in a very primitive form, later *Gêrfrid* (Förstemann, 1). The loss of the first of the two *r*'s through dissimilation is natural ; so is the development of *ā*, represented by modern [ou]; cp. O.E. *gāt > goat.*

GRANBY.

Type I.

1086 Granebi, D.B.
1284 Granbi ⎫ F.A.
1302 Graneby ⎭

Type II.

1086 ⎫ Grenebi ⎰ D.B.
1184 ⎭ ⎱ Index.

"The *bȳ(r)* or farm of *Grani*." The pers. n. forming the
first element comes from the Scandinavian language; so does
the ending. The spelling under Type II is due to an attempt
at connecting the name with the adjective O.E. *grēne*, "green."

GRASSTHORPE or GREISTHORPE [gresþōp].

1086 } Grestorp { D.B.
1268 }   { Index.

1302 Gresthorp }
1316 Grethorp } F.A.

1424 } Gresthorp { Index.
1637 }   { Map in Camden.

All the old spellings point to M.E. *gres þorp*, "grass thorpe
or village." The form *gres* is explained by Dr Björkman (Scand.
Loanwords, p. 30) as due to Norse influence. This tallies
with the fact that the second element too is derived from the
Scandinavian.

I can make nothing of the second modern spelling: it is
hardly likely that the Scandinavian male name *Grīs, Grȳse*
(Björkman, s.v.) has had any influence. The first modern form
shows influence of Standard English *grass*, which has, however,
not yet affected the local pronunciation.

GREASLEY [grīzli].

Type I.

1086 Griseleia, D.B.
1428 Grisseley, F.A.

Type II.

1216–72 Greselley, Inq. P.M. I.
1284 Gresley   }
1302 Gresseley } F.A.
1346 Greseley }

Type III.

1284 Grelley, F.A.

Type IV.

1637 Graseley, Map in Camden.

The original meaning is preserved by Type I : "the *lēah* or field of *Gris*." The latter name of Scand. descent is treated by Dr Björkman. Type II, which seems to have survived eventually, although I cannot explain the quantity of the first vowel satisfactorily, arose out of the confusion with M.E. *gres*, "grass" (see *Grassthorpe*). Type III is derived from Type II, *sl* [zl] having become assimilated to *ll*. In Type IV the *e* is replaced by *a* on the analogy of the modern standard form *grass* (or *graze*).

GREET (river).

958 (ondlang) greotan, Cart. Sax. 1029.

The above quotation from a charter is of very questionable value, and I am inclined to set aside its evidence altogether, as far as the final *n* is concerned. I take this name to represent O.E. *grēot ēa*, or *ēʒe*, "gravelly, sandy river," from O.E. *grēot*, "gravel, sand," and *ēa*, *ēʒe*, "water, water-course." There exists another river Greet in Worcestershire. The termination has disappeared completely as in the original river-name *Blyth* (q.v.).

On O.H.G. territory a corresponding formation *Griezpah* is found (Förstemann, II).

GRIMSTON.

1086 Grimestune, D.B.
1302 Grymeston, F.A.

From O.E. *Grīmes tūn*, "the farm of *Grīm(r)*." The pers. n. involved is Scandinavian in origin. The long *ī* was shortened before the combination *mst*.

GRINGLEY (Little) (Grinley, Hope).

Type I.

(a)   1086 $\left\{ \begin{array}{l} \text{Grenleige} \\ \text{Grenelei} \end{array} \right\}$ D.B.

1278 Grenlay $\left\{ \begin{array}{l} \text{H.R.} \\ \text{Inq. P.M. II.} \end{array} \right.$

1316 Grenleye, F.A.

1327–77 Grenley, Non. Inq.

(*b*) 1375 Grynley, Index.

Type II.

c. 1300 Gringelay, Index.

Type III.

1704 Little Grimley, Map.

O.E. *on þǣre grēnan lēage*, "on the green field or plain." The neighbourhood is noted for its meadows. Clareborough, which derives its name from "clover," is close by.

O.E. *ē* was shortened during the M.E. period after having been raised to *ī* (Type I *b*). Type II arose out of confusion with the following name. The genuine etymology is preserved in the modern local pronunciation, as recorded by Hope, but not in the official spelling. The change from *in* to *ing* is remarkable; it is probably due to confusion with the following name.

The inventor of Type III fancifully connected the name with the pers. n. *Grim* contained in Grimston, Grimsby etc.

There is a place called Grindley in Staffordshire which has the same origin and meaning. The *d* arose out of a phonetic glide between *n* and *l*.

GRINGLEY-ON-THE-HILL.

Type I.

1086 Gringeleia, D.B.

c. 1200 Gringhelaya, Cal. Rot. Chart.

1278 Gringele, H.R.

1316 Gringeley, F.A.

1327–77 Gryngeley, Non. Inq.

1372 Grynglay, Index.

Type II.

1086 Greneleig (?), D.B.

1535 { Greynley
Grenely } of the Hill } Valor Eccles.
Grynley on the Hill

Type II most probably arose out of confusion with the preceding name. As a matter of fact, it is extremely difficult, if not at times impossible, to distinguish between the two in early records. In the Inq. P.M. c. 1500, e.g., both are hopelessly mixed up. The D.B. spelling quoted under Type II may refer to the preceding place.

I take the first element to be a variant of the O.N. pers. n. *Grīmketell*, *Grīmkell*, *Grinkell*. The existence of a type *Gringel* is conclusively proved by Prof. Wyld, who refers to the above pl. n. and further adduces *Grimgelege* and *Gringelthorp* (Pl. Ns. of Lancs., s.v. *Cringlebarrow Wood*, p. 102).

It is highly interesting to trace the development of that personal name through the succeeding stages of shortenings and assimilations. First the *ī* is shortened before the combination *mk*, and the last syllable loses its vowel: *Grīmketell* > *Grimketll*. Then *t* is assimilated to *l*, *m* to *k*: *Griŋkel(l)*. After that the *k* is voiced under the influence of the surrounding sounds; result, *Gringel*.

The meaning is " *Grimkel's* field or meadow."

GROVE.

$$\left.\begin{array}{l} 1086 \\ 1216\text{--}1307 \\ 1302 \end{array}\right\} \text{Grave} \left\{\begin{array}{l} \text{D.B.} \\ \text{Testa de N.} \\ \text{F.A.} \end{array}\right.$$

From O.E. *æt þǣm grāfe*, "at the grove." Taking into account the modern spelling, this is the only explanation I can offer, although the very late persistence of the *a* in the above forms might speak against that derivation, and in favour of O.E. *æt þǣm græfe*, "at the grave, or sepulchral mound." It is, however, possible that this persistence is only apparent, the spelling being copied from earlier records.

GUNTHORPE [gɑnþɔ̄p].

Type I.

1086 Gulnetorp, D.B.
? Gunildethorp, Thoroton, III, 25.

Type II.

1086 Gunnetorp, D.B.
1278 Guntorp, H.R.
1302 Gunthorp, F.A.
1489 Gownthorpe, Woll. MSS.

If the spelling quoted by Thoroton as an ancient one is really genuine and was, as is not unlikely, taken by him from old local records, the etymology of the name is clear : *Gunnhildar þorp*, " the village of *Gunhildr*." The latter is a feminine pers. n., Norse in origin, and rather frequent in M.E. times. It is very rare that places are called after women. The D.B. spelling under Type I may very well be an attempt at representing a pronunciation **Gunelthorp < Gunildarþorp*. It is just what one would expect the compiler of D.B. to do. The havoc played by Anglo-Norman scribes among the liquids and nasals of English place-names is well illustrated by numerous examples collected in Dr Zachrisson's book (pp. 120 sqq.). Metathesis of *dl* (> *ld*) is found in several instances in D.B. (Stolze, § 30).

If the above explanation of Type I is accepted, Type II is but a further shortening of the original. If it is rejected as based upon doubtful or spurious evidence, the name may be taken to contain as first element the O N. masculine pers. n. *Gunner*.

It may be mentioned here that the occurrence of *Gunhild* in a pl. n. is better attested for the Yorks. Gunthwaite, which is spelt *Gunnyldthwayt* in a document of 1389 (Descriptive Catalogue of Ancient Deeds, I, Index).

HABBLESTHORPE, see APPESTHORPE.

HAGGONFIELD (under Worksop).

The absence of early spellings makes it impossible to suggest an etymology with any degree of certainty.

HALLAM [heiləm].

> 1331 Halum ⎫
> 1541 Halom ⎬ Index.
> 1853 Halom, White, Directory.

O.E. *æt healum*, " at the valleys." The place is situated at the foot of a lofty range of hills. The ending represents the dat. pl. of O.E. *healh*. All the early forms contain but one *l*, so that the first syllable was open, which explains the quantity of the vowel.

Isaac Taylor (Engl. Village Names) explains this name as meaning "at the slopes." The exact sense of O.E. *healh* is very uncertain; the transition from "valley" to "slope" is easy and natural.

HALLOUGHTON [hōtn] (Hortn, Hope).

$\left.\begin{array}{l}1291 \\ 1428\end{array}\right\}$ Halton $\left\{\begin{array}{l}\text{Tax. Eccl.} \\ \text{F.A.}\end{array}\right.$

1637 Haulton, Map in Camden.

O.E. *healh tūn*, "the farmstead in the *healh* or valley." We have to deal with two types, of which the first and most primitive is, curiously enough, represented by the modern spelling only. The latter may rest on local tradition strengthened by records not generally available.

The O.E. prototype has developed in two directions. Type I: A glide arose between *l* and *h* similar to that in *borough* from O.E. *burh, buruh* and spelt in the same fashion. This accounts for the modern official form. Type II: The *h* was dropped between the *l* and the *t*; afterwards an *u*-glide developed before the *l* which then disappeared. The old spellings and the modern pronunciation illustrate the latter line of change.

HARBY.

1086 $\left\{\begin{array}{l}\text{Herdebi} \\ \text{Herdrebi}\end{array}\right\}$ D.B.

1291 Herdebi, Tax. Eccles.

1302 Hordeby, F.A.

The additional *r* in the second D.B. form is due to a blunder. The name stands for O.E. *heorda bȳ(r)*, "the herdsmen's dwelling," from Mercian *heorde*, "a herd," and Scand. *bȳr*, "dwelling." The *o* in the F.A. spelling is a M.E. representative of older *eo*. The phonetic development of the name is regular: *d* is lost between two consonants, *er* becomes *ar* (Phonology, § 8).

HARWELL or HAREWELL.

1086 Herewelle, D.B.

1227-77 Herewell, Non. Inq.

The etymology of this name is uncertain. Various suggestions of a more or less convincing nature can be offered.

(1) The first element looks like O.E. *here*, "army, band of thieves," in a special sense "the Danish army." It is not unlikely that this spring (*well*) took its name from the fact that an army of the Norsemen once camped near it on one of the numerous plundering expeditions mentioned in the Anglo-Saxon Chronicle. If this suggestion is accepted, the meaning of the pl. n. would be *æt þǣre here welle*, "at the spring or brook of the Danish army." Similar names are found on the continent. Förstemann (II) records O.H.G. *Heribrunnum, Heriburnon.*

(2) The adjective *heoru-weallende*, "fiercely boiling," occurs in O.E. poetry and might very well be applied to a spring or brook. A similar compound *heoru-well*, "the fiercely boiling spring," is a possible ancestor of the modern name. There is nothing extraordinary in the suggestion of a fierce and raging well or stream of water. Several brooks called *Wuodaha* in O.H.G., modern *Wutach*[1], meaning "raging brook," are found in Germany.

The phonetic development is regular. The alternative spelling represents an attempt at popular etymology.

HARWORTH [hærəþ] (Harroth, Hope).

    1086 Hạreworde, D.B.
    1278 Harewurh, H.R.
    1346 Hareworth, F.A.

From O.E. *Hearanweorþ*, "the homestead of *Heara*." The pers. n. involved is recorded once in the Onomasticon. The final *h* in the H.R. spelling is an unsuccessful attempt by the Norman scribe to render the unfamiliar spirant *þ*.

HATFIELD (under Norton).

    1278 Haytfeld, H.R.
    1291 Hattefeld, Cal. Rot. Chart.

---

[1] See Schröder, p. 8. I do not agree with the interpretation of this name as "producing rage, or madness." What the idea underlying the creation of this name was, may be gathered from the following quotation from Grabbe's "Hermannschlacht" (Zweiter Tag): "Helft doch unserer armen Retlage. Sie wollen den Bach überschreiten, und so klein er ist, wehrt er sich und *schwillt ganz ärgerlich auf*!"

1332 $\begin{cases} \text{Hethfeld} \\ \text{Hadfeld} \\ \text{Hatfeld} \end{cases}$ For. Rec. ed. Stevenson, p. 399.

1571 Hatfield, Index.

O.E. *hǣþ féld*, "the heathy field, or plain." The H.R. spelling distinctly points to influence of Scand. *heiðr*, "heath." The *a* of the modern form is the result of early shortening of O.E. *ǣ* before the combination *tf*. The stopping of the open *þ* before *f* is a curious feature. Mr Duignan gives the same explanation for the Staffs. *Hatfield*. The pl. n. spelt *Hæðfeld* in Bede is modern *Hatfield*.

Hatfield is now the name of two farms in Norton, but originally it was applied to the district in the north of the county marked by sand and fen. It is, therefore, clear that the O.E. *hǣþ féld* was first employed to designate the whole plain, and that the village which sprang up in it became known as *in hǣþ félde*, "in the heath field."

HAUGHTON.

Type I.

1086 Hoctun, D.B.

1203 $\begin{cases} \text{Hoctune} \\ \text{Octon} \end{cases}$ Index.

1278 Hockton, H.R.

Type II.

1375 Hoghton, Index.

"The *tūn* or farm of *Hōc*." Although there is no sign of a genitive termination, I take the first element to be the Scandinavian man's name *Hauk(r)*, anglicised to *Hōc*. The phonetic changes resulting in the modern form are in accordance with general rules. O.E. *Hōk(r)tūn* (perhaps containing the Scand. genitival *r*, see Björkman, p. 184) > *Hoktun* with shortening of *ō* before *kt* (Phonology, § 1); this is Type I. Before *t*, *k* was opened, developing into *h* (Phonology, § 20), a change which led to Type II and eventually to the modern form. On *au* for M.E. *ou* see Phonology, § 9.

M. 5

HAWKSWORTH.

Type I.

1189 Houkeswarda, P.R.

Type II.

1086 Hochesuorde, D.B.

1302 $\begin{cases} \text{Hokeworth} \\ \text{Hokesworth} \end{cases}$ F.A.

"The homestead of *Hauk(r)* or *Hōc.*" Type I and the present-day spelling exhibit the Scandinavian pers. n. in its more primitive shape. The diphthong *ou* was anglicised to *ō* which is found in Type II. On modern *au* for M.E. *ou* see Phonology, § 9.

HAWTON.

Type I.

1086 Holtone, D.B.

Type II.

1086 Houtune, D.B.

$\left. \begin{matrix} 1227-77 \\ 1291 \\ 1302 \end{matrix} \right\}$ Houton $\begin{cases} \text{Non. Inq.} \\ \text{Tax. Eccles.} \\ \text{F.A.} \end{cases}$

Type III.

1270 Hautone, Index.

O.E. *holt-tūn,* "the dwelling in the wood," from O.E. *holt,* "holt, wood." The name is identical in meaning with the numerous *Woottons* < *wudu tūn.* In Type I, the original *l* is preserved; between it and the preceding *o,* an *u*-glide arose. The *l* then disappeared (see Phonology, § 9). In M.E., the *ou* of Type II soon developed into *au,* giving rise to Type III which is represented by the modern spelling (see Phonology, § 9).

HAYTON.

1154–89 Haythona, Index.

$\left. \begin{matrix} 1278 \\ 1327-77 \end{matrix} \right\}$ Hayton $\begin{cases} \text{H.R.} \\ \text{Non. Inq.} \end{cases}$

1428 Heyton, F.A.

Probably *heið tūn,* "the farm in the heath." The first element is Scand. *heið(r),* the cognate of O.E. *hǣþ,* "heath."

HAYWOOD OAKS.

Haywood stands for M.E. *heȝe wood*, "a fenced in or enclosed wood"; O.E. *heȝe*, a derivative of O.E. *haga*, means "a hedge, fence."

HAZELFORD FERRY [hæzlfəd].

1278 Hesilford, H.R.

The meaning is obvious. In northern M.E., the name of the plant involved was *hesel*, from Scand. *hesli*, "a collection of hazels." See Rygh, N. Gaardnavne, p. 57. The latter form is found in the spelling of 1278. This probably is the correct form, whereas the modern name has been influenced by the Standard English variant *hazel*. In White's Directory (1853), the name is spelt "Heaselford," *ea* representing M.E. *ę̄*, from older *e* in open syllable.

HEADON [hīdn].

1086 Hedune, D.B.
1302 Hedun, F.A.
1362 Hedon, Index.

O.E. *æt þǣre hēan dūne*, "at the high hill." The *n* of the weak dative must have been lost at a very early period (Sweet, N.E. Gr. §§ 1030–33). O.E. *dūn*, "hill," is the ancestor of modern *down*, "open high land."

HECK DYKE (brook).

The second part is O.E. *dīc*, "ditch," see *Bycardyke*. I cannot explain the first element.

HEDGROVE (in Southwell).

No early forms. It is possible that the name stands for *hedge grove* [hedȝ grouv], with loss of the middle consonant [ȝ]; cf. Phonology, § 12.

HEMPSHILL [hemsəl].

Type I.
1086 Hamessel, D.B.
1275 Hamdisel ⎫
1278 Homeshull ⎭ H.R.

Type II.

(*a*) c. 1200 Hemdeshill, Woll. MSS.

1216–1307 Hemdeshil, Testa de N.

1275 Hemdeshyll, H.R.

(*b*) 1209–10 Hindeshull, P.R.

(*c*) 1702 Hempsall, Index.

The first element has the appearance of a man's name in the genitive case. The second seems to stand for either O.E. (W. Sax.) *setl*, (Northern) *seðel, séld*, " seat, abode, residence," or O.E. *hyll*, " hill." The spellings under Type II *b* and *c* are influenced by analogy of *hind*, "female deer," and of *hemp*, the plant, respectively.

HESLEY.

1217 Heselay, Index.

" The hazel lea, or open field." The first element is Scandinavian *hesli*, " hazel grove," see *Hazelford*.

HICKLING [iklin̩].

1086 $\left\{ \begin{array}{l} \text{Hechelinge} \\ \text{Hegelinge} \end{array} \right\}$ D.B.

1284 Hickelinge, F.A.

1291 Hicling, Tax. Eccles.

An O.E. patronymic : *æt (H)iclingum*, " at the dwelling-place of the family of *Hicel*." The *Iclingas* were a noble family to whom St Guthlac belonged. It is, however, by no means certain that Hickling was a settlement of that particular clan. The descendants of any man called *Hicel* would be styled *Hicelingas*.

The *e* in the D.B. forms stands for O.E. *i* according to a frequent practice of Anglo-Norman scribes (see Stolze, § 9).

HOCKERTON.

1086 $\left\{ \begin{array}{l} \text{Hocretune} \\ \text{Ocreton} \end{array} \right\}$ D.B.

c. 1200 Hocretona, Index.

1302 Hokyrton, F.A.

" The *tūn* or homestead of *Hōc*." This pers. n. is Scandinavian in origin (see *Hawksworth*). The *r* represents the Norse genitival ending (see Björkman, p. 184).

HODSOCK.

Type I.

1086 Odesach, D.B.
1302 Hodesak
1316 Hoddisack ⎱ F.A.
1346 Hodelake (*sic!*) ⎰

Type II.

1278 Hoddeshock, Inq. P.M. II.

The second element of this pl. n. is O.E. *āc*, "oak"; the first is a pers. n., probably Scandinavian *Oddi*. The *h* may be due to influence of the O.E. pers. ns. *Hod, Hoda, Hodo* (Onomasticon); it is more likely, however, that it is a mere inorganic addition, initial *h* being a very unstable element in this dialect (see Phonology, § 19).

The vowel of O.E. *āc* was shortened at two different periods in the unstressed syllable: (1) In O.E. times; result *a*, Type I. (2) In M.E.; result *o* (< *ō*), Type II and modern spelling. The second *h* of Type II is due to confusion with *hōc*, "a heel, promontory."

N.B. It may be mentioned here that *Hodesac* is found in an O.E. charter (Cart. Sax. 1282) among Worcestershire field names and boundaries.

HOLBECK or HOWBECK or HOLBEACH.

1329 Holbeck, Index.

"The brook in the hollow." From O.E. *holh*, "hollow," and Scand. *bekk(r)*, "a brook." The termination *-beach* goes back to O.E. *bæc*, with palatalised *c*, the native equivalent of *bekkr*. This word is found in pl. ns. ending in *batch, bach, beach*, and is discussed at length by Professor Skeat in his Pl. Ns. of Cambs. (pp. 44 sq.).

It is impossible to say whether the O.E. or Scand. form was the original one.

The vocalisation of *l* is treated elsewhere (Phonology, § 9).

HOLME.

c. 1200 Olm, Index.

1316 Holme, F.A.

This common English pl. n. is derived from Scand. *holm(r)*. Its meaning is "island in river, land rising from water." There is an open pasture on the Trent bank at Normanton called "the Holme." The above pl. n. seems to have had the same meaning originally.

HOLME PIERREPONT.

1086 Holmo, D.B.
1302 Holm, F.A.

For the meaning of M.E. *holm* see the preceding name. The final *o* in D.B. is remarkable; I do not pretend to be able to explain it.

The distinctive addition is due to the fact, that the *Pierrepont* family of Norman descent, now the Earls Manvers, owned and still own the manor: "Annora de *Perpunt* tenet manerium de *Holm*," F.A. 1302.

HORSEPOOL (now the name of a decayed farm near Thurgarton).

1086 Horspol, D.B.
1302 Horppol (*sic*!) ⎫
1316 Horspol   ⎬ F.A.
                   ⎭

The name explains itself.

HOUGHTON, see HAUGHTON.

HOVERINGHAM.

Type I.

(*a*) 1278 Offringham, H.R.
       1428 Overyngham, F.A.
       1535–43 Oringgam, Leland.
(*b*) 1086 Horingeham, D.B.
       1278 Hofrungham, H.R.
       1316 Horingham, F.A.

Type II.

1346 Heveringham ⎫
1428 Heveryngham ⎬ F.A.
                  ⎭

From O.E. *Eoforinga hām*, "the home or village of the family of *Eofor*." Type II shows the regular descendant of O.E. *eo*, which is *e*. The presence of *o* in Type I is probably the result of a peculiar development of O.E. *eo* before a labial (*v*) in the dialect. The Scand. man's name *Iofr*, the equivalent of *Eofor*—which, by the way, means " boar "—may have influenced the first element, just as it did in the pl. n. *York* (see Place Names of the W. Riding, p. i).

The addition of an inorganic initial *h* (Type I *b*, Type II) cannot surprise us (see Phonology, § 19).

The same patronymic name occurs on the continent. *Ebringen* (Breisgau) goes back to O.H.G. *Eburingen*, and *Everghem*, near Ghent, is derived from O. Low Germ. *Everinge-hēm* (Förstemann, I and II).

HUCKNALL TORKARD.

1086 $\left\{\begin{array}{l}\text{Hochenale}\\\text{Hochehale}\end{array}\right\}$ D.B.

1190 Hukenhall, Woll. MSS.

1216–1307 Hukenhall, Testa de N.

1284 Hukenall

1302 Huckenale Torkard $\left.\right\}$ F.A.

1316 Hokenall

1327–77 Hukenhale, Inq. P.M. II.

O.E. *æt Huc(c)an heale*, " at or in the valley of *Hucca*." This pers. n. is recorded in the Onomasticon as *Huc* and *Hucco*. It seems to be a short or " pet " form of a full name beginning with *Hyge-*.

The distinctive addition owes its origin to the fact that the manor was once held by the Norman family of Torkard: " Johannes *Torkard* tenet in *Hukenall*," F.A. 1284.

HUCKNALL under HUTHWAITE or DIRTY HUCKNALL [dɑti(h)ʌknə].

c. 1500 Durty Hucknall, Inq. P.M. c. 1500.

1611 Hucknoll Huthwaite, Index.

1704 Dirty Hucknall. Quoted by Horner Groves, Hist. of Mansfield, from parish register (p. 170).

See preceding and following names. The flattering addition of *dirty* (M.E. *dritig*, from Scandinavian, see Björkman, Scand. Loan Words) is probably due to the former condition of the

roads and the surrounding country. I am told that whereas
Hucknall-T. stands on limestone, which absorbs water quickly,
this village is situated chiefly on clay.

HUTHWAITE [haþweit], now the official name of the place
above.

Probably a late settlement, as is shown by the Scand. origin
of the second element, *þveit*, " piece of land, an outlying cottage
with its paddock " (Vigf.). The first element may be the pers. n.
taken from the neighbouring Hucknall : *Hucþwait > Huthwaite*
by assimilation of *kþ* to *þ(þ)*.

IDLE (river).

> 627 Idla, Beda, Hist. Eccles.
> 1302 Yddil, Index.

The second element may be O.E. *ēa*, *ēʒe*, "water, stream,"
which was lost entirely as in *Greet*, *Blyth* (q.v.). If that is so,
I would suggest that the first element was O.E. *īdel*. The
recorded meanings of this adjective—" empty, desolate, useless "
—do not seem to be applicable to a river. It is, however,
believed by some scholars that the original sense was " shining,
brilliant," it being related to Greek *αἴθω* (see Kluge's and
Hirt's Etym. Dictionaries, s.v. *eitel*). If that is so, the sense of
O.E. *sēo īdele ēa* would be " the bright, clear river."

IDLETON, see EATON.

KELHAM.

> 1086 Calun, D.B.
> 1189 ⎫
> 1225 ⎭ Kelum ⎰ P.R.
>   ⎱ Bor. Rec.
> 1227–77 Kelm, Non. Inq.
> 1244 ⎫
> 1302 ⎭ Kelum ⎰ Index.
>   ⎱ F.A.
> 1316 Kelme, F.A.
> 1350 Kelom ⎫
> 1453 Kelum ⎬ Index.
> 1578 Kellam ⎭

From O.E. *æt celdum*, " at the water-courses." This name
corresponds to the Latin *aquis*. The initial *k*-sound proves that
the noun involved is derived from the Scandinavian language,

O.N. *kelda*, "well, spring, brook flowing from a spring." The modern spelling in -*ham* is due to false analogy with pl. ns. ending in O.E. *hām*, "home."

The same meaning attaches to the neighbouring *Averham* (q.v.).

KERSALL.

Type I.

1086 Cherveshale, D.B.

Type II.

1302 Kyrneshall ⎫
1316 Kyrnissale ⎭ F.A.

The second element is O.E. *healh* (dative *heale*), "a valley." The pers. n. contained in the first element cannot be determined with accuracy. The *v* in D.B. may be a misreading for *u*, which in turn might stand for *n*; if that be so, Type I would be the same as Type II. The latter may contain the O.E. male name *Crin* or *Crina*, recorded in the Onomasticon. Metathesis of *r* is frequently met with in this dialect (Phonology, § 15). *er* in the modern form is a mere spelling variant for *ir*, both representing the same sound in present-day English.

KEYWORTH [locally: kjuəþ; otherwise: kīwʌ̄þ].

Type I.

1086 Caworde, D.B.

Type II.

1200 Kyeword, Cal. Rot. Chart.

1216–1307 Kewurch, Testa de N.

c. 1294 Keword, Woll. MSS.

1302 ⎧ Kewrth ⎫
⎨ Kewrht ⎬ F.A.
⎩ Keworth ⎭

1637 Kyworth, Map in Camden.

The D.B. scribe seems to have mistaken the first element for the pers. n. *Cawa, Ceawa*. I take this name to stand for O.E. *cȳworþ*, "the cow enclosure, or farm." *cȳ* represents either the genitive singular or the nominative plural of O.E. *cū*, "cow." The spellings show that by the year 1200 the accent has been shifted from the $\bar{y}$ to the u developed out of the w. After this change, the $\bar{y}$ ($\bar{\imath}$) degenerated into a mere palatal glide [j]. With these sound-changes may be compared the analogous history of modern English *ewe* < O.E. *ēowu*. The combination *ey* in the modern form is a mere spelling device to express this glide, which is represented by *e* or *y* in earlier records. The polite pronunciation is based entirely on the written form, whereas locally the etymologically correct form survives.

KILVINGTON.

1086 $\left\{\begin{array}{l}\text{Chilvintun} \\ \text{Chelvinctone}\end{array}\right\}$ D.B.

1291 Kilington, Tax. Eccl.

1302 Kylvington $\left.\right\}$
1428 Kylyngton $\left.\right\}$ F.A.

1637 Skillington, Map in Camden

From O.E. *Cylfinga tūn*, "the *tūn* or homestead of *Cylfa's* descendants." The *c* in the second D.B. spelling clearly demonstrates that a patronymic name forms the first element. *Cylfa* is once recorded in the Onomasticon. From a number of early forms it would appear that the *v* was dropped at an early period in pronunciation but retained in writing. From this fact one would expect the local pronunciation to be [killintn, killiŋtn]. The initial *S* of Camden's map must be due to a blunder.

KIMBERLEY.

1086 $\left\{\begin{array}{l}\text{Chinemareleie} \\ \text{Chinemarelie}\end{array}\right\}$ D.B.

c. 1200 Kinemarle, Woll. MSS.

1227–77 Kymm'ley, Non. Inq.

1291 Kynmarley, Tax. Eccles.

1316 Kynmarleye $\left.\right\}$
1428 Kymerley $\left.\right\}$ F.A.

1589 Kymmerley *alias* Kymberley, Index.

O.E. *Cynemǣres lēah*, "the field or open country of *Cynemǣr*." It is remarkable that the genitival *s* is absent from all the early spellings. The phonetic history can be easily traced through the centuries, and affords instructive examples of the various processes of shortening and assimilation to which pl. ns. are liable.

KINGSTON-UPON-SOAR.

Type I.

1086 Chinestan, D.B.

$\left.\begin{array}{l}\text{1216–72}\\\text{1256}\end{array}\right\}$ Kyn(n)estan $\left\{\begin{array}{l}\text{Index.}\\\text{Inq. P.M. I.}\end{array}\right.$

1291 Kynstan, Tax. Eccles.

$\left.\begin{array}{l}\text{1302 Kyneston}\\\text{1346 Kynston}\end{array}\right\}$ F.A.

1637 Kynston, Map in Camden.

Type II.

1592 Kyngston, Index.

O.E. *cyne-stān*, "royal stone." The first element is the O.E. adj. *cyne-*, "royal," which occurs in compounds. I am unable to say what the "royal stone" involved was, or why it was so named. It must be remembered that in O.E. times, the title of "king" was applied not only to the rulers of large dominions but also to the petty chiefs of minor clans; hence its frequent use as a modern surname. In fact, the word originally had the wide meaning of "nobleman, one of a noble family" (see Kluge, Etymol. Wörterbuch, s.v. *König*).

The change from *a* to *o* in the ending is not due to phonetic development but to erroneous etymology, the *s* being considered the genitival ending of some pers. n. preceding the more familiar *ton*, O.E. *tūn*. Thoroton the historian must have been labouring under the same delusion when he explains the name as, "So called, probably, from an Owner, as most Towns of that termination, in this County, generally."

Popular etymology is also responsible for the change from Type I to Type II.

KINOULTON.

1086 Chineltune, D.B.

1152 Cheneldestoa, Index.

$\left.\begin{array}{l}\text{1284 Kynalton}\\\text{1302 Kynolton}\end{array}\right\}$ F.A.

O.E. *Cynewealdes tūn*, "the farmstead of *Cyneweald*." This is an example of the complete loss of the genitival *s* which is preserved only in the most archaic form. The phonetic development is regular (see Phonology, §§ 18 ; 9).

The final *a* for *n* of the Index is, of course, a scribal mistake.

KIRKBY-IN-ASHFIELD.

1086 Chirchebi, D.B.

$$1240 \left\{ \begin{array}{l} \text{Kirkeby} \\ \text{Kirkby} \\ \text{Kyrkeby} \end{array} \right\} \text{Bor. Rec.}$$

$$\left. \begin{array}{l} 1302 \quad \text{Kirkeby} \\ 1316 \quad \text{Kirkeby super Asshefeld} \end{array} \right\} \text{F.A.}$$

A Norse name, of which many instances are found both in England and in Scandinavia : " the church village," from O.N. *kirkja* and *bȳ(r)*. It may be remarked here that the combination *ch* in D.B. does not express the sound of *ch* in modern English *church*, but represents *k* before palatal vowels (Stolze, § 40, 1, 2).

The distinctive addition refers to the district in which the village is situated : O.E. *æsc féld*, " the field or plain of the ash-tree(s)."

KIRKLINGTON.

Type I.

1086 Cherlington, D.B.

1291 Kirtelyngton, Tax. Eccles.

$$\left. \begin{array}{l} 1302 \quad \text{Kyrtelington} \\ 1428 \quad \text{Kyrtelynton} \end{array} \right\} \text{F.A}$$

c. 1500 Kyrtelyngton, Inq. P.M. c. 1500.

Type II.

1346 Kyrkelington, F.A.

$$\left. \begin{array}{l} 1437 \\ \text{c. } 1500 \end{array} \right\} \text{Kirklington} \left\{ \begin{array}{l} \text{Index.} \\ \text{Inq. P.M. c. 1500.} \end{array} \right.$$

" The dwelling-place of the family of *Cyrtel*," O.E. *Cyrtelinga tūn*. The pers. n. involved is discussed under *Costock*. It is formed from a name *Curt* by addition of the diminutive ending *il* which caused mutation of the preceding vowel.

The change from *tl* to *kl* may be due to purely phonetic forces. However, the analogy of the independent word *kirk* may have had some influence in rendering Type II the predominant form.

The Yorks. pl. n. Kertlington is spelt *Kirtlington* and *Kirklington* in Cal. Inq. P.M. temp. Edw. I. It seems to contain the same patronymic.

KIRTON.

Type I.

1086 Circeton, D.B.

Type II.

1086 Chircheton, D.B.

c. 1200 } Kyrketon { Index.
1316 } { F.A.

c. 1500 { Kyrkton } Inq. P.M. c. 1500.
{ Kyrton }

"The church town, or village." Type I is entirely English in character, whereas Type II substitutes Scand. *kirkja* for the cognate Anglo-Saxon *čiriče*. In D.B. *c* before palatal vowels is used to render the sound of modern English *ch* [tʃ], whereas *ch* in the same position stands for *k* (see Stolze, § 40, I, and cp. *Kirkby*).

KNAPTHORPE (under Caunton).

1086 Chenapetorp, D.B.

1278 Konaptorp, H.R.

1302 Knapthorp, F.A.

"The hamlet of *Knapp*, or *Knappi*." This pers. n. is not given by Dr Björkman. However, the second element of the pl. n. *þorp* being of Scand. origin, it is safe to conclude that the first too comes from the same source. Rygh (G. Pers. Nav. p. 161) gives examples of the occurrence of *Knappr*, *Knappi* in O.N. pl. ns.

It might also be suggested that this name contained the O.E. substantive *cnæp(þ)*, "top, mountain top," M.E. *knape*[1]. I am, however, strongly in favour of the first interpretation.

The vowels *e, o* between the *k* and *n* were put in by the Normans who found it impossible to pronounce the two consonants in this combination.

[1] Leland says of Belvoir Castle that it "standith on the very *Knape* of an highe Hille."

KNEESALL [nīsə] (Kneaser, Hope).

1086 Cheneshale, D.B.
1189 Cneeshala, P.R.
1230 Keneshale, Index.
1278 Kneshale, H.R.
1302 Kneshall ⎫
1316 Kneshale ⎭ F.A.

The second element is O.E. *healh*, "valley." I cannot explain the first, although I suspect it to stand for O.E. *Cnihtes*, or *cnihtes*, the genitive of either the pers. n. *Cniht*, or the identical noun meaning "boy, servant, attendant, retainer." If it could be proved that the genuine development of O.E. *ih* in this dialect was late M.E. *ī*, this theory would receive considerable support. The Dial. Grammar is too untrustworthy a guide to be relied upon in such cases. I have been informed that old people in the district say [nīt] for *night*, but have never been able to establish this fact beyond doubt. The voiceless *s* in the pronunciation recorded above would seem to point back to early M.E. *ts* (cp. *Cossal*).

The vowel-glide between the *k* and the *n* in Norman documents is explained elsewhere (see *Knapthorpe*; and the following name).

KNEETON or KNEVETON.

Type I.

1695 Knighton, Map in Camden.

Type II.

1086 Chenivetone, D.B.
c. 1190 Chnivetun ⎫
c. 1210 Knītona ⎭ Woll. MSS.
1284 Knyveton, F.A.
1291 Kenyveton, Tax. Eccles.

O.E. *cnihta tūn*, "the farmstead, or settlement of the servants." The exact sense in which the O.E. *cniht* is used here is obscure. Originally it meant "servant," as its German cognate *Knecht* still does. In course of time, the O.E. word assumed a different meaning, being applied to the retainers of a king or powerful lord whom they served as warriors, or men-at-arms. They belonged to a new nobility, ranking above the lower orders from which they had sprung. The word is, however, most certainly not used here in its still further specialised M.E. sense, that of "chevalier." Isaac Taylor (Engl. Village Names, § 3) counts fifteen villages called *Knighton* in England.

On modern *ee* [ī] for O.E. *ih* see preceding name.

The *v* of Type II remains to be explained. I take it to represent the faint palatal open consonant of the early M.E. *Knighton*, as it appeared to the Normans who were unfamiliar with that sound (see Zachrisson, pp. 119 sqq.).

Considering that the mediæval documents were often written by foreigners, and were moreover copied one from another, it is not surprising that so late an authority as Camden should furnish us with the most useful spelling.

LAMBLEY.
>1086 Lambeleia, D.B.
>1316 Lameleye, F.A.

" Lamb lea, or field." O.E. *lamb lēah.*

LAMCOTE.
>*Type I.*
>1086 Lanbecotes, D.B.
>*Type II.*
>1086 Lanbecote, D.B.
>1316 Lambecote, F.A.

O.E. *lamb cotas,* "the lamb cotes" (Type I), *æt lamb cotum,* "at the lamb cotes" (Type II); *mb* is often spelt *nb* in D.B.; see early forms of *Clumber,* and *Cromwell.*

LANEHAM.
>1086 Lanun, D.B.
>1278 } Lanum } H.R.
>1316 } ..... { F.A.

O.E. *æt lanum,* "at the lanes," dat. pl. of *lane,* "lane, street, narrow way between hedges or banks." The modern spelling is another instance of the O.E. dat. pl. ending *-um* being mistaken for *hām,* "home."

LANGAR.

$$\left.\begin{matrix} 1086 \\ \text{c. }1190 \end{matrix}\right\} \text{Langare} \left\{\begin{matrix} \text{D.B.} \\ \text{Woll. MSS.} \end{matrix}\right.$$

1241 Langar, Cal. Rot. Chart.

$$\left.\begin{matrix} 1302 \\ 1637 \end{matrix}\right\} \text{Langer} \left\{\begin{matrix} \text{F.A.} \\ \text{Map in Camden.} \end{matrix}\right.$$

O.E. *se langagār*, "the long triangular strip of land." The second element *gār* corresponds to the O.H.G. *gêro*, modern German *Gehre*, which is encountered in numerous field names throughout Northern Germany. It is discussed at length by Dr Jellinghaus (p. 283). The modern Engl. equivalent is *gore*, "a triangular piece of land."

LANGFORD.

1086 Landeforde, D.B.

$$\left.\begin{matrix} 1302 \\ 1346 \\ 1428 \end{matrix}\right\} \text{Landeford, F.A.}$$

$$\left.\begin{matrix} 1470 \ \text{Lanford} \\ 1472 \ \text{Lanforth} \\ 1555 \ \text{Landford} \end{matrix}\right\} \text{Index.}$$

"The ford leading to the plough-lands." The change of *n* (< *nd*) to *ng* [ŋ] is very remarkable. It did not take place until a comparatively late period. The dialect word *lang*, "long," may have had some influence.

O.E. *land* has the special sense of "cultivated land, estate." There was a pl. n. called *Lanfurt* in Friesland (Förstemann, II) which contained the same element, *land*, meaning "ager, rus."

The spelling in *-forth* (1472) betrays influence of Scand. *fiorðr*.

LANGWITH.

1291 Langwaith, Cal. Rot. Chart.

1571 Langwith, Index.

It is impossible to say whether the second element stands for Scand. *vað*, "a wading place, ford," or for *viðr*, "tree, wood, forest." See other examples of this confusion under Vað and Viðr, pp. 394 and 395 of Lancs. Pl. Ns. However, I feel inclined to explain the name as meaning "long ford." For such a name there are several analogies: Langwith Wood, Yorks., appears as "Haya (an enclosed wood) de *Langwath*" in 1286 (Index), and there is a *Longford* in Lancs.

LAXTON.

Type I.

(*a*)   1086   Laxintune, D.B.
      1278   Lexington, H.R.
  c. 1300   Lexinton, Index.
(*b*)   1278   Lessinton, H.R.

Type II.

$\left.\begin{array}{l}1291\\1302\end{array}\right\}$ Laxton $\left\{\begin{array}{l}\text{Tax. Eccles.}\\\text{H.R.}\end{array}\right.$

"The *tūn* or homestead of *Leaxa*, or the *Leaxings*." This pers. n. is found once in O.E., in the Index to Kemble's Cod. Dipl. as first element of a pl. n.: *Leaxan oc*, "the oak of *Leaxa*." The phonetic development can be traced through the early spellings without difficulty. The *ss* < *ks* of Type I *b* distinctly points to Norman influence.

LEAKE (East or Great L., and West or Little L.).

1086 $\left\{\begin{array}{l}\text{Lecche}\\\text{Leche}\end{array}\right\}$ D.B.

1227–77 Leyk, Non. Inq.

1302 $\left\{\begin{array}{l}\text{Esterlek}\\\text{Westerleke}\end{array}\right\}$ F.A.

1637 $\left\{\begin{array}{l}\text{Esterleak}\\\text{Westerleak}\end{array}\right\}$ Map in Camden.

"At the brook." The name is derived from an O.N. word *løkr*, "a brook, rivulet," which is a cognate of O.E. *lacu*, "lake, running water," and connected with modern Engl. *to leak*. The village of East Leake is situated in a hollow of the South Wolds through which a small rivulet flows.

LEEN (river).

    c. 1200 Liene, Woll. MSS.
    1535–43 Line Ryver, Leland, I, 103.
    1637 Lin, Camden, p. 547.

M.   6

Apparently a Celtic river-name. There is a river *Len* in Kent, and we also find a *Leenane* in co. Galway, a *Leane* in co. Kerry, and a *Leanane* in co. Donegal. See Stevenson, Asser's Life of King Alfred, p. 318, where a similar river-name is discussed.

The name *Lenbach* is found on Bavarian territory.

LENTON.

> 1086 Lentune, D.B.
> 1189 }
> 1291 } Lenton { Nott'm Charter.
>   { Index.
> 1637 { Linton }
>   { Lenton } Camden, p. 547.

"The *tūn*, or settlement on the river Leen." See *Leen*.

LEVERTON (North and South).

Type I.

1086 Cledretone, D.B.

Type II.

(*a*) 1086 Legretune, D.B.
    c. 1200 Legherton, Cal. Rot. Chart.
    1278 { Nordleg'ton }
        { Sudleg'ton } H.R.
(*b*) 1189 Leirton, P.R.
    1216–1307 Leyrton, Testa de N.
    1281 Norhtleyrton, Inq. P.M. I.

Type III.

1173–4 Leuerton, P.R.

The etymology of this name is very obscure. Type III, which may be the most reliable spelling, looks as if it were derived from an O.E. compound *Lēofhere(s) tūn*. The same personal name seems to be contained in "Liverpool," see Pl. Ns. of Lancs.

For a possible explanation of the relations between *d* [ð], *g*, and *v*, in Norman spelling see discussion under *Averham*.

LIDE (more correctly *Lythe*) (Wapentake, now part of the hundred of Thurgarton).

Type I.

1086 Lide Wap, D.B.

1302 ⎫
1346 ⎭ Wapentakium de...Lith, F.A.

Lythe, Thoroton.

Type II.

1278 Lye, H.R.

(Thurgarton a) Lee, Thoroton.

This name seems to be that of a river. Perhaps from O.E. *līþe ēa*, "the gentle stream." *līþe* is identical with modern Germ. *linde*, both meaning "mild, calm, gentle, pleasant." The adjective was occasionally applied to flowing water, as in a M.E. version of the Psalter, quoted in N.E.D., s.v. *lithe*: "His stremes leften lithe" (et siluerunt fluctus ejus, cvi. 29).

See *Blyth*, which has the same meaning.

The loss of intervocalic ð (Type II) is a regular feature.

LIMPOOL (under Harworth).

c. 1500 Lympole, Inq. P.M. c. 1500.

Probably from O.E. *æt līnd pōle*, "at the pool of the lime-trees."

LINBY.

1086 Lidebi, D.B.

1316 Lyndebi, F.A.

O.E. *līnd bȳ(r)*, "the dwelling near the lime-trees." From O.E. *līnd*, "lime-tree," and Scand. *bȳ(r)*, "a dwelling, village."

The D.B. scribes frequently omit *n* before other dental consonants (Stolze, § 32); this is due to a peculiarity of their Norman-French pronunciation: the *i* in the above spelling was nasalised, and the consonant *n* dropped.

LITTLEBOROUGH.

Type I.

1535 Litilborowe, Valor Eccles.

Type II.

1086 Litelburg, D.B.

1428 Lytilburgh, F.A.

1535 Litilbrugh, Valor Eccles.

1637 Lyttleburgh, Map in Camden.

"The small fortified place, the small Roman fort." The word O.E. *burh* is often applied to old fortifications especially of Roman origin. It is constantly used of Rome itself, of London and other walled cities. Littleborough is generally believed to stand on or near the site of the Roman station Agelocum or Segelocum. Cp. *Brough.*

Type II represents the O.E. nominative *lȳtlo bur(u)h,* whereas Type I stands for the dative *æt þǣre lȳtlan burȝe,* with *u* for *y* by analogy of the nom. and accus. cases, Sievers, § 284, anm. 4.

LOUND.

$$\left.\begin{matrix}1086\\1278\end{matrix}\right\} \text{Lund} \left\{\begin{matrix}\text{D.B.}\\\text{H.R.}\end{matrix}\right.$$

1302 Lound, F.A.

"The wood, copse." From Scand. *lundr,* "wood, copse." The lengthening of the *u* before *nd* took place in England after the word was introduced into the language. This word enters into the composition of numerous pl. ns. both in Scandinavia and in the Norse districts of Great Britain. It is also found as an independent word in various forms in the English dialects.

The name of the university town of Lund in Southern Sweden is identical in origin.

LOWDHAM [laudm].

1086 Ludeham, D.B.

$$\left.\begin{matrix}\text{c. 1170 Ludam}\\\text{c. 1200 Ludham}\end{matrix}\right\} \text{Woll. MSS.}$$

$$\left.\begin{matrix}1278\\1291\end{matrix}\right\} \text{Ludham} \left\{\begin{matrix}\text{H.R.}\\\text{Tax. Eccles.}\end{matrix}\right.$$

1302 Loudham, F.A.

1637 Lewdham, Map in Camden.

This name cannot be explained with any degree of certainty. The first element may contain the O.E. pers. n. *Lūda,* which seems to be an abbreviated variant of a full name beginning with *Lūd*, *Lēod-*, such as *Lūdhere*, *Lēodmǣr* etc. *Lūdan ham* would mean "the homestead of Lūda."

Camden's spelling looks very suspicious. It is most probably a mere blunder.

It might also be suggested that the first element was O.E. *hlūd*, "loud," which I take to mean "stormy, windy" as well: cp. O.E. *hlȳda*, a name for the month of March, derived from *hlūd*, "windy," and therefore identical with the *ventôse* of the French revolutionary calendar. "The stormy-homestead" seems to possess a parallel in the O.H.G. *Hludinhusir* (Förstemann) which may mean "at the stormy houses," but the first element may equally well be a pers. n. in the genitive case. Cp. *Bleasby*.

LYNDHURST-ON-THE-FOREST.

1637 Lyndhurst Wood, Map in Camden.

Formerly a wood and part of the forest. The meaning is obvious: "Lime-wood," from O.E. *lind*, "lime-tree," and *hyrst*, "a wood, copse, grove."

MANSFIELD.

Type I.

A.

(*a*) 1086 Mamesfelde, D.B.

1163 Mammesfeld, Index.

1291 } Mamesfeld { Tax. Eccles.
1316 } { F.A.

(*b*) 1291 Mannesfeld, Tax. Eccles.

B.

(*a*) 1189 Mamefeld, P.R.

(*b*) 1278 Man'efeld, H.R.

Type II.

A.

(*a*) 1086 Mammesfed, D.B.

(*b*) 1227–77 Maunnesfeld, Non. Inq.

1278 Maunsfewd, H.R.

1428 Maunsfeld, F.A.

1564 Mawnsfeld, Index.

1657 Maunsfeld, Map in Camden.

B.

(*a*) 1249 Malmefeud, Inq. P.M. I.

The O.E. prototype of this name was *Mammes feld* (Types I A and II A), of which a variant *Mamman feld* (Types I B and II B) existed. These O.E. prototypes of which, as will appear later, the latter was probably the more original one, are most faithfully preserved by the Index spelling of 1163, and the P.R. of 1189 respectively.

The name developed on two different lines among the English and the Norman-French communities. Type I, the native form, survived eventually in the pl. n.; Type II, characterised by the development of a *u* between *a* and *n* (Phonology, § 11), and the vocalisation or disappearance of *l* (cp. Zachrisson, pp. 146 sqq.)[1], owes its origin to the peculiarities of Norman-French pronunciation. From the latter type, the name of the river on which Mansfield stands is taken; it is therefore wrong to say that the town derives its name from the river, just the opposite being the case (see *Maun*).

In both types we perceive the change from medial *m* to *n*; those forms marked (*a*) contain the former, those marked (*b*) the latter consonant. This development may be due to several causes acting simultaneously.

(1) Dissimilation of the sequence *m—m—f*; two consecutive *m*'s followed by *f* are difficult to pronounce. This applies especially to Norman-French speakers (see Zachrisson, pp. 120 sqq.).

(2) The vowel between the *m* and the *s* must have disappeared very early, at least in pronunciation if not in spelling. In the combination *ms*, *s* would exercise a very strong assimilatory influence upon the preceding nasal.

(3) Popular etymology connected the first element with the word *man*.

Having explained the development of the name through M.E. and modern times as exemplified by the variety of early spellings, we may now return to the original O.E. form. The meaning of

---

[1] *l* having become *u* in Anglo-Norman, *al* could be written for *au* as in *Malmefeud* (Type II B, *a*).

Mammes féld is obvious : " the plain of *Mamma*." This name applied to the whole district, the town, or rather the original settlement being called " *on Mammes félde*." The final *e* of the D.B. form (Type I A, *a*) may be regarded as the last and only trace of the O.E. dative ending. As has already been said *Mamman feld* was probably more original than the *-es* type. The pers. n. *Mamma* should follow the weak declension, and must have done so originally ; however, as was shown in the case of *Annesley* (q.v.), the ending *an* was replaced by the strong *es* at a very early date in the Midland dialects.

The pers. n. involved is not recorded in O.E. sources. It may, however, be safely inferred to have existed, because it occurs among those West Germanic tribes which remained on the continent. We find O.H.G. *Mamo, Mammo* as the names of persons, and *Mammindorf*, modern *Mammendorf*, Bavaria, as the name of a place (Förstemann, I).

Who the personage was, that gave his name to the plain and town, it is impossible to say. It even is not unlikely that he never existed except in the imagination of the early settlers. The locality may have had a British name, which contained the element *Mam-*, of doubtful meaning and derivation, which is also found in the early forms of *Manchester*. This *Mam-*, whose significance the Anglo-Saxons did not know, would promptly be interpreted and used as the pers. n. with which they were already acquainted. They did this the more readily, as the majority of Teutonic pl. ns. were formed on the principle of pers. n. plus designation of locality (see Bradley, Engl. Histor. Review, Oct. 1911, p. 823). Similar cases of distortion and misinterpretation of British pl. ns. are cited by Isaac Taylor (Words and Places, ch. XII).

Apart from the erroneous, but natural derivation of this name from that of the river, another different and highly ingenious explanation has been offered, namely, that it was bestowed upon the locality by the Counts of Mansfeld in Saxony who came here to attend at the tournament of King Arthur. It may seem a pity to many that the hard facts should destroy so romantic a fiction.

MANTON.

1086 Mennetune, D.B.

O.E. *Mǣnan tūn*, "the farmstead of *Mǣna*." This pers. n. is recorded in the Onomasticon. The long vowel was shortened before the combination *nt* at different periods, which accounts for the variation in vowel of the only two recorded spellings given above.

MAPELBECK.

Type I.

1086 Mapelbec, D.B.
c. 1300 Mapilbec, Index.
1302 Mapelbek ⎱ F.A.
1316 Mapulbek ⎰
c. 1500 { Mapulbeke ⎱ Inq. P.M. c. 1500.
{ Malebeke ⎰

Type II.

1086 { Mapleberg ⎱ D.B.
{ Mapelberg ⎰

" The maple brook." From O.E. *mapul(trēo)*, " maple (tree)," and O.N. *bekkr*, " brook, rivulet" (Type I). The small watercourse on which the village stands is now without an individual name.

Type II contains O.E. *beorg*, "hill, mountain," as second element. It is impossible to say whether this represents the original name which was later changed to the prevailing form. Mapelbeck is, as a matter of fact, situated at the foot of a hill which might very well have been called "the maple hill." It must, however, be mentioned here that the compilers of D.B. are by no means trustworthy guides in matters of etymology, as they too frequently employed the expedient of substituting more familiar forms for those with which they were not acquainted (see *Bramcote*).

MAPPLEWELL.

" The maple spring." The place probably takes its name from a spring in that locality, which is also famous for the petrifying qualities of its very cold and pure water.

MARKHAM (East, or Great).

1086 Marcham, D.B.

MARKHAM (West).

1086 Westmarcham, D.B.

From O.E. *mearc hām*, "the home or dwelling on the boundary." The meaning of O.E. *mearc* was "boundary, mark, district"; it refers to the boundaries of states, but more frequently of fields and estates. The word *mearc land* was used to describe the waste land which often formed the boundaries of extensive clearings, and it is not impossible that the original Markham was situated on the confines of such a district. A political boundary may also have been implied, an assumption which is rendered likely by the fact that East Markham is situated on the watershed ridge of the Trent and Idle, and thus on a natural boundary which might easily have become a political one.

MARNHAM.

Type I.

1086 Marneham, D.B.

c. 1175 Marnaham, Woll. MSS.

1302 Marnhame, F.A.

Type II.

c. 1190 Marisham { Coucher Book of Walley Abbey, Cheetham Soc., 1847, p. 5.

I take this name to stand for O.E. *Mǣrwines hām*, "the homestead of *Mǣrwine*," a pers. n. recorded in the Onomasticon. Type II represents the more natural development of the proto-type, with the *ǣ* shortened before *rw*, and subsequent loss of initial *w* and *n* before *s* in the unstressed syllable. In Type I, which survived, the same changes took place with the one exception that the *s* instead of the *n* was dropped. It is very likely that the two types go back to two forms of different length, and therefore, stress, viz. the nominative and dative respectively: *Mǣrwines hām*, and *æt Mǣrwines hāme*, though what the exact distribution of accents was I am at present unable to suggest (see Wyld, in Pl. Ns. of Oxf. pp. 5 sqq.).

MARTIN or MORTON.

> 1086 Martune, D.B.
> 1217 Marton ⎱
> 1216–72 Martun ⎰ Index.
> c. 1500 Marton, Inq. P.M. c. 1500.

"The boundary farm or enclosure." The first element may have been either O.E. *mǣre*, "boundary," or *mearc* with the same meaning. Although the *k* would readily disappear between *r* and *t* (Phonology, § 12), it is safer to adopt the former alternative in the absence of spellings containing a *c* or *k*. The word *mǣre* is used exclusively of the boundaries of estates and fields. (Lancs. Pl. Ns. p. 370.) What was the nature of the boundary here referred to, it is impossible to say. See *Markham*.

Both modern spellings are due to popular etymology: the first owes its existence to the analogy of the Christian name Martin, the second to that of the numerous pl. ns. Morton. There is, however, some phonetic justification for the former, as the unstressed vowel after dental consonants, and more especially if followed by another dental, is often pronounced *i* in English dialects (Horn, § 149, 2 *a*).

Martin forms the north-eastern hamlet of Harworth parish, adjoining Bawtry, which is situated in Yorkshire. The boundary between the two counties seems thus to go back to a very old division.

MATTERSEY or MATTERSEA.

Type I.

> 1086 Madressei(g), D.B.
> 1278 ⎱   ⎧ H.R.
> 1316 ⎬ Mathersey ⎨ F.A.
> 1428 ⎰   ⎩ F.A.
> c. 1500 Mathersey, Inq. P.M. c. 1500.

Type II.

> 1291 Marsey, Tax. Eccles.
> 1335 Mersey, Valor Eccles.

Type III.

(*a*) c. 1500 Madersay, Inq. P.M. c. 1500.
  1535–43 Madersey, Leland.
(*b*) c. 1500 Matteseythorp, Inq. P.M. c. 1500.
  1535 Matersey, Valor Eccles.

"The island or low-lying water meadow of *Mæðhere.*" The same personal name is contained in the Worcestershire pl. n. *Madresfield.* The second element is O.E. *ēʒe.*

Type I preserves the original most faithfully. The spellings under Type II show the regular loss of *ð* between vowels. If this form had survived one would expect the modern pronunciation to be [māzǐ]. Type III, which accounts for the modern spelling, arose out of the stopping of *ð* before *r* (III *a*); the *d* thus produced was unvoiced during a later period, probably from a desire to vary the nature of the sounds which in this long word were all voiced.

The ending *sea* in the alternative modern spelling is an ingenious attempt at etymology.

The nature of the ground in the neighbourhood, as described by Leland, seems to be in accord with the meaning of the suffix. He says: "...and a Mile farther I saw the Course on the lifte hond of...Ryver, over the which I passid by a Bridge of...hard at the entering into *Madersey* Village. Thens I rood a Myle yn low wasch and sumwhat fenny Ground...."

MAUN or MAN (river).
  1300 Main(esheued)* ⎫ Stevenson, Forest
  1332 Mamm(esheued)*⎬ Records, 399, 401.
  1613 Man, Drayton's Polyolbion.
  c. 1900 Man, or Maun, Ordnance Map.
    * *i.e.* head of river Maun.

The name of this river is derived from that of the town of *Mansfield.* Similar "back-formations" are enumerated by Dr Bradley in Essays and Studies, I, pp. 32, 33[1]. The variations

---

[1] "Thus Kimbolton, in Huntingdonshire, is derived from the personal name Cynebald ; but the river on which the place stands has been provided by the map-makers with the name Kim....The name of the river Brain is a figment invented to account for Braintree."

in spelling are explained under *Mansfield* (q.v.); *ain* for *an* is a peculiarly central French development; see *Saundby*.

MEDEN (river) [mīdn].

The termination may be the reduced form of an O.E. *amma*, or a similar word for "stream"; see *Witham*. I propose to connect the first part with O.E. *mǣd*, "meadow." The meaning "meadow stream" seems a most appropriate one for this particular water-course. A corresponding O.H.G. *Madibah* is recorded several times (Förstemann, II).

MEERING.

Type I.

1086 Meringe, D.B.
1302 Meryng, F.A.

Type II.

1216–1307 $\left\{ \begin{array}{l} \text{Meringes} \\ \text{Meriges} \end{array} \right\}$ Testa de N.

This is a patronymic name, which is proved by the forms of Type II. I believe this to have been O.E. (non-West-Saxon) *Mērwingas*, "the family or tribe of the *Mērwings*" (Type II), of which Type I is the dative pl. *æt Mērwingum*.

This name is identical with that of the noble family of the Merovings, who as a dynasty preceded the Carolings in France and North-Western Germany. The original pers. n., of which this is the patronymic, must have been W. Germ. *māru*, "bright, famous" (see Hirt-Weigand, Deutsches Wörterb., s.v. *Märe*).

The *Meringas* are mentioned in an O.E. charter (Cod. Dipl. 809), and the same patronymic occurs in the O.H.G. pl. n. *Maringen*, modern *Möhringen* (Förstemann, II). The name *Maring, Mering* is frequent in the O.H.G. period, and is represented by the modern German surname *Mehring*.

Isaac Taylor, in his work entitled "Words and Places," refers to the Merovingians in connection with the above pl. n. and gives numerous references (ch. VII).

MERRILS BRIDGE (West Drayton).

1225 de ponte Miriild }
1316 Mirielbrigge } Bor. Rec.

"The bridge by the pleasant slope," O.E. *sēo myrige helde*, from *myrig*, modern "merry," here used in its old and original sense of "pleasant, delightful" as in "Merry England," and *helde*, West-Saxon *hielde*, "slope, declivity." The O.E. *y* of the first syllable is represented by *i* in the M.E. forms. The *e* in the modern spelling may be the result of a peculiar dialect development of O.E. *y*, or may be due to the influence of the independent word, *merry*, which comes from the Kentish dialect.

For an explanation of the final *s* see *Brentshill*.

Merrils Bridge is a very ancient structure situated at the foot of a gentle slope.

MIDDLETHORPE (under Caunton).

c. 1500 Midelthorp, Inq. P.M. c. 1500.
1704 Midlethorp, Map.

The name explains itself.

MILNTHORPE (under Norton).

Probably from M.E. *milen thorp*, "the mill thorpe or hamlet." The O.E. word for a mill is *mylen*. See *Milton*.

MILTON or MILNETON.

Type I.

1278 Milneton, H.R.
c. 1500 { Milnton }
         { Mylneton } Inq. P.M. c. 1500.

Type II.
1086 Miletune, D.B.

Type III.
c. 1500 Molton, Inq. P.M. c. 1500.

O.E. *mylen tūn*, "the enclosure, or farmstead by the mill." See *Milnthorpe*.

Types II and III show loss or assimilation of *n* between *l* and *t*. The *o* of Type III, which probably stands for *u*, may be due to a particular M.E. development of O.E. *y* after labials.

MIRFIELD HALL.

I take *Mirfield* to stand for O.E. *se myrige féld*, " the pleasant plain, or field." See *Merrils Bridge* for old meaning of *myrige*.

MISSON (Mizon, Hope).

1086 $\left\{ \begin{matrix} \text{Misna} \\ \text{Misne} \end{matrix} \right\}$ D.B.

1278 Misin, H.R.

1321 Mysyn $\left. \begin{matrix} \\ \\ \end{matrix} \right\}$ Index.
1379 Misen

1637 Masson, Map in Camden.

I cannot interpret this name. It seems to contain a river-name as first element; the *a* and *e* of D.B. seem to stand for older *ēa*, " river," so that the prototype would be O.E. *æt M...ēa*, " (the habitation) by the M...river." This *ēa* or *a* would be lost subsequently as in *Blyth* (q.v.).

Camden seems to have blundered.

I cannot refrain from quoting the following delightful interpretation of the name contained in White's Directory (1853, p. 640): " Misson Parish lies...on the north side of the Idle, bounded on the west by Yorkshire, and on the east by Lincolnshire, and is partly in the latter county, which is here so intermixed with Nottinghamshire, that the boundaries of the two counties are almost indefinable, from which circumstance the parish is supposed to have been anciently called *Misne* or *Myssen*."

MISTERTON.

1086 Ministretone, D.B.

1278 $\left\{ \begin{matrix} \text{Misterton} \\ \text{Mist'ton} \end{matrix} \right\}$ H.R.

1316 Misterton, F.A.

O.E. *mynster tūn*, " the minster-town, or habitation by the church." The meaning of O.E. *mynster* is " monastery, nunnery, church, cathedral"; in this case it probably refers to an old pre-Norman structure. The loss of *n* after initial *m* may be due to dissimilation (Phonology, § 14).

Misterton in Leicestershire has the same origin; it appears as *Minstertona* in 1313 (Index).

MORTON or MORETON (under Babworth).

$$1086 \left\{ \begin{array}{l} \text{Mortune} \\ \text{Nordermortune} \end{array} \right\} \text{D.B.}$$

$$\begin{array}{l} \text{c. 1200} \\ 1316 \end{array} \right\} \text{Morton} \left\{ \begin{array}{l} \text{Index.} \\ \text{F.A.} \end{array} \right.$$

MORTON (Fiskerton-with-Morton).

$$\begin{array}{l} 1331 \\ 1368 \\ 1754 \end{array} \right\} \text{Morton, Index.}$$

The meaning is obvious: "the *tūn* or habitation on the moor." The distinction made in D.B. seems to have been lost subsequently.

MUSKHAM (North and South).

$$1086 \left\{ \begin{array}{l} \text{Nordmuscham} \\ \text{Nord Muscham} \end{array} \right\} \text{D.B.}$$

$$\begin{array}{l} 1143 \\ 1189 \end{array} \right\} \text{Muscamp} \left\{ \begin{array}{l} \text{Index.} \\ \text{P.R.} \end{array} \right.$$

1316 Suthe Muskham, F.A.

1637 Muscombs, Map in Camden.

O.E. *Muscan hām*, "the home or dwelling-place of *Musca*." This pers. n. is not recorded in the Onomasticon. It must, however, have existed as it is found on the continent; Förstemann (1) quotes from German documents *Musco*, *Musgo*, and refers to the modern German surname *Musch*. There is a pl. n. in Hesse exactly identical with the one under discussion: *Muschenheim*, from O.H.G. *Muscanheim*.

The same pers. n. seems to be contained in the Scandinavian pl. ns. *Muskedalen*, *Muskerφd* which are left unexplained by Rygh (N. Gaardnavne, p. 375). If the English name is not altogether of Norse origin, the retention of the pronunciation *sk* must at any rate be ascribed to Scandinavian influence.

Camden again presents us with a fanciful spelling; he probably thought the name was connected with *combe*. The final *s* is the sign of the plural, there being two villages of the same name.

The Index and P.R. forms betray Norman influence: the

ending was taken to represent *camp*, the Norman-French descendant of Latin *campus*, frequently found in pl. ns. such as *Fécamp*.

NETHERFIELD.

" The lower field."

NETTLEWORTH (under Warsop).

> 1216–1307 Nettelwurd, Testa de N.
> c. 1500 Nettilworth, Inq. P.M. c. 1500.
> 1637 Nettleworth, Map in Camden.

" The enclosed homestead, habitation among the nettles "? There is no evidence either to prove or disprove this interpretation conclusively.

NEWARK-UPON-TRENT.

> 1066 Newarcha, Cod. Dipl. 878 (a starred charter).
> 1086 $\begin{cases} \text{Newerca} \\ \text{Neuuerce} \\ \text{Newerche} \end{cases}$ D.B.
> 1189 $\begin{cases} \text{Niwewerch, P.R.} \\ \text{Niwerch, Nott'm Charter.} \end{cases}$
> 1278 Newerk, H.R.
> etc.

O.E. *þæt niuwe weorc*, "the new fortification." The old fortifications, probably a continuation of Roman works, were destroyed by the Danes, but rebuilt in the reign of Edward the Confessor, when the place received its present name.

The meaning of the O.E. *weorc*, "fortification," is still preserved in the compound *earth-work*, and in the special military sense of "the works." Cp. also "out-works." The modern German *Werk* has the same sense.

For *er* > *ar* see Phonology, § 7.

NEWBOLD (under Kinoulton).

Type I.

> $\left.\begin{array}{l} 1086 \\ 1284 \end{array}\right\}$ Neubold $\left\{\begin{array}{l} \text{D.B.} \\ \text{F.A.} \end{array}\right.$

Type II.

1157 Niwebote }
1159 Niwebota } P.R.

O.E. *þæt niuwe botl*, or *bold*, "the new dwelling." The noun appears in both forms *botl* and *bold* in O.E. (Sievers, § 183, 2 *a*), which accounts for the difference between Types I and II.

NEWINGTON (under Misson).

O.E. *æt þǣm niuwan tūne*, "at the new homestead." This name is found in all parts of England. For the change of unstressed *an* to *ing* see Phonology, § 13.

The same name in the nominative case *sē niuwa tūn* is the prototype of the equally numerous Newtons.

NEWSTEAD ABBEY.

1189 (Prior de) Novo Loco, Nott'm Charter.
1205 Novus Locus in Shirewood, Cal. Rot. Chart.

c. 1500 { New Place } Inq. P.M. c. 1500.
{ Newstede }
{ Newstead }

The Priory, less correctly, Abbey of Newstead was founded in 1170 by Henry II in what was then part of Sherwood Forest. Most probably the site had to be cleared of trees and undergrowth and was therefore called *Novus Locus* by the monks, though it is impossible to say whether this was a translation of a M.E. *Newe stede* or whether the latter was based on the original Latin name. The *New Place* of 1500 certainly looks as if it were a translation from the Latin; this variant may have been the form adopted by the Norman-French among clergy and nobility.

NEWTHORPE.

1086 Neutorp, D.B.
1341 Neuthorp, Index.

This name requires no explanation.

M.   7

NEWTON.

1086 Niwetune, D.B.

c. 1250 Neuton, Index.

O.E. *sē niuwa tūn*, "the new homestead," see *Newington*.

NORMANTON-ON-SOAR.

1086 $\left\{ \begin{array}{l} \text{Normantune} \\ \text{Normanton} \end{array} \right\}$ D.B.

NORMANTON-ON-TRENT.

1086 Normentune, D.B.

1268 Normanthon, Index.

NORMANTON-ON-THE-WOLDS.

1086 Normantone, D.B.

O.E. *norðmanna tūn*, "the enclosure or dwelling-place of the Northmen." These places owe their names not to the Normans but to their non-Frenchified kinsmen, the Scandinavian invaders. It might be urged against this that the suffix *tūn* is Anglo-Saxon and not Norse in origin. We must, however, bear in mind that the pl. n. was invented and used by the original Anglo-Saxon inhabitants among whom the newcomers made their settlements; pl. ns. combining a Scandinavian pers. n. with an Anglo-Saxon designation of locality must be viewed in the same light.

The loss of *ð* between *r* and *m* is normal.

NORNEY (under Styrrup).

This place, of which there are no early spellings, is situated to the north of Blyth on a small river. I take it to stand for O.E. *norðerne ēʒe*, "the Northern brook, or island." The development would be as follows : *norðrneʒe > norðnei > Norney*.

NORTON (in Cuckney Parish).

1282 Norton, Bodl. Ch. and R.

"The north town, or habitation." This village forms the northern part of Cuckney.

NORWELL.

1086 Nortwelle, D.B.

1278 { Nortwell } H.R.
       { Norwell }

1316 Northewelle } F.A.
1428 Northwell   }

"The northern spring or brook."

NOTOWN (under Bleasby).

As there are no early spellings it is impossible to attempt an explanation.

NOTTINGHAM [notigəm, Dial. Gramm. § 273] (Nottingum, Nottingyum, Nottinum, Hope).

Type I.

868 etc. } Snotingaham { A.S. Chron. *passim.*
930   }   { Cod. Dipl. II, 170.

1086 { Snotingham } D.B.
     { Snotingeham }

c. 1150 Snotingaham { Flor. of Worcester.
                    { Symeon of Durham.

c. 1250 Snotingham, MS. Jesus Coll. Oxon. (E.E.T.Soc. 49).

1353 Snotyngham, Leicester Records.

Type II.

1131 } Notingeham { P.R.
1153 }   { Index.

1278 Notingham, H.R.

c. 1300 Notingeham, Henry of Huntingdon.

1304 } Notingham { Index.
1336 }   { Leicester Records.

N.B. The majority of the early spellings quoted above are taken from Dr Zachrisson's book on "Anglo-Norman Influence on English Pl. Ns." (pp. 51, 52).

O.E. *Snotinga hām,* "the homestead of the family of *Snot.*" The pers. n. *Snot* occurs as that of a tenant in D.B. The name *Snothere* is also recorded in the Onomasticon. There may have existed an O.H.G. pers. n. *Snozo* which seems to be contained in the pl. n. *Snozindorf* (Förstemann, I). The weak form of the O.E. name, *Snotta*, survives in M.E. *Snotte*, the surname of a certain Peter mentioned in the Close Rolls (Cal. of Close Rolls, p. 570, quoted by Mr Stevenson).

Type I represents the native Anglo-Saxon form; Type II, with the initial *S* dropped, owes its origin to Norman-French influence. Romance-speaking peoples find great difficulty in pronouncing certain initial consonant combinations of the Germanic languages. When a word was borrowed, such combinations were naturally got rid of, either by prefixing an *e*, or by inserting some vowel between the two consonants, or by simply dropping the obnoxious initial sound. The latter alternative was adopted when the Normans[1] had to use the name of the O.E. *Snotinga hām*, and that of the neighbouring Sneinton, which has a similar origin, being derived from O.E. *Snotinga tūn* (see D.B. spellings under *Sneinton*). The fact that the clipped form survived in the former case only is accounted for by the circumstance that Nottingham, with its castle, became a most important stronghold of the conquerors, who settled in such numbers in the town that it had to be divided into two distinct communities, an English and a French one (see Zachrisson, pp. 51 sqq.).

There still exists a general belief, even among people that ought to know better, that the name of the "Queen of the Midlands" signifies "the home of the caves." But however romantic and appropriate this interpretation[2] may be, it will have to be abandoned. The notion is taken from a passage in Asser's "Life of King Alfred" (ed. W. H. Stevenson, M.A., Oxford, 1904, p. 230), which reads: "Snotengaham...quod Britannice 'Tigguobauc' interpretatur, Latine autem 'speluncarum domus.'" The learned editor of the text remarks that the

---

[1] Dr Bradley (E. and St. p. 39) facetiously remarks "that the people of Nottingham will bear them no ill-will on this account."

[2] I cannot refrain from quoting the following delightful explanation of the name given with the utmost assurance by Mrs Gilbert in her pamphlet entitled "Recollections of Old Nottingham" (p. 7): "Snottengham, from Snottenga (caves) and ham (home), subsequently softened into Nottingham."

British name actually does mean "dwelling of the caves" or, more literally, "cavy house"; but this has nothing to do with the English form, which is quite a new creation. This interpretation was later on eagerly seized upon by antiquaries, who made it the basis of fanciful elaborations in which they delighted, being concerned more with grotesque fiction than with sober facts. Camden in particular must be credited with having amplified and widely circulated the original mistake of King Alfred's biographer. The passage is so quaint and characteristic that it may find a place here: "Where, on the other banke (of the *Lin*) at the very meeting well neere of *Lin* and *Trent*, the principall Towne that hath given name unto the Shire is seated upon the side of an hill now called *Nottingham* (by softning the old name a little) for *Snottengaham*; for, so the English Saxons named it of certaine caves and passages under the ground, which in old time they hewed and wrought hollow under those huge and steepe cliffs, which are on the South side hanging over the little River *Lin*, for places of receit and refuge, yea and for habitations. And thereupon *Asserius* interpreteth the Saxon word Sottengaham in Latine *Speluncarum domum*, that is, *An house of Dennes or Caves*, and in the British *Tuiogobauc*, which signifieth the very selfe same" (Camden, 547).

NUTHALL or NUTTALL [nɑtl].

> 1086 Nutehale, D.B.
> 1284 Notehall ⎫
> 1302 Notehale ⎪
> 1316 Notehall ⎬ F.A.
> 1428 Notehale ⎭

O.E. *on hnutu hĕale*, "in the nut valley, in the vale where the nuts grow." Similar names, as *hnut fen*, *hnut wic*, are quoted by Middendorf from O.E. charters. This is one of the few names that show early substitution of *hall* for the second element. The principal modern spelling owes its origin to the same erroneous conception of the meaning of the ending.

The *o* of all the F.A. forms stands for M.E. *u* according to Norman-French practice (Sweet, N.E. Gr. § 775).

OLDCOATES or ULCOATES (Alecotes, Hope).

Type I.

(*a*) 1302 ⎫
       1346 ⎬ Oulecotes, F.A.

1348   Oullecotes, Index.

1428   Oullecotez, F.A.

1445   Owelcotes, Index.

(*b*) 1269   Ulcotes, Cal. Rot. Chart.

1278   Ulecotes, H.R.

1414   Ulcotes, Index.

1535   Ulcotts, Valor Eccles.

Type II.

1086   Caldecotes, D.B.

O.E. *ūlan cotas*, "the houses of the owl, or near which the owl lives, is seen or heard." The vowel of the first syllable remains long in Type I *a*, *ou* or *ow* being the M.E. (Norman-French) symbols denoting the long *u*-sound. This vowel is shortened before the combination *lk* in Type I *b*; from the latter the second modern spelling originates. The principal official form owes its origin to the interference of popular etymology. At some time or other the dialect pronunciation of the two words *owl* and *old* may have been very much alike, and may have caused the substitution of the latter for the former. The only information given by the Dial. Dict. is that M.E. *ū* becomes [au] or [ā] in Nottinghamshire, whereas *old* is pronounced [ōd]. The transcription of the local pronunciation furnished by Hope is very ambiguous; if it is interpreted in accordance with the ordinary principles of modern English spelling, it would mean [eilkouts] or [ēlkōts]. In spite of inquiries instituted in the locality itself a pronunciation deviating from the spelling could not be traced.

The D.B. form is without support. The scribe seems to have substituted a name with which he was more familiar. There is a *Coldcotes* in the West Riding. Dr Moorman interprets this as meaning "the cold cottages, on an exposed situation." Isaac Taylor (Words and Places, ch. X) is of opinion, that this name like that of *Cold Harbour* (i.e. *auberge*) was given to certain structures erected on frequented roads, where travellers could obtain shelter but neither food nor fire[1]. When such a name belongs to a place not in the immediate neighbourhood of an ancient road, it seems to me to mean not "cottage on an exposed position," but "temporary building, house without a fire-place," such as one may still find in the fields used as barns and temporary shelters for cattle.

The Hessian pl. n. *Eudorf* is explained by Sturmfels (p. 21) as meaning "Dorf, wo sich der Uhu gerne aufhält"—"village where the owl delights to dwell"; cp. O.H.G. *hûwo, hûo*; *ûwila*, "owl."

OLDWARK SPRING.

Stukeley, the antiquarian, found extensive Roman remains near this place. The meaning is therefore: "the spring near the old work or buildings." See *Newark*.

OLLERTON or ALLERTON.

1086 Alretun, D.B.
1189–99 Alretona ⎫Index.
  1190   Alretun ⎭
1278 Alverton, H.R.
1316 ⎫   ⎧ F.A.
1377 ⎬ Allerton ⎨ Index.
1637 ⎭   ⎩ Map in Camden.

Probably from O.E. *Ælfheres tūn*, "the enclosure or homestead of *Ælfhere*." The H.R. spelling of 1278 is the most valuable. It does not, however, enable us to say for certain that the pers. n. contained in the first element is *Ælfhere* rather than either *Ælfred* or *Ælfric* or *Ælfweard*.

The change from *al* to *aul > ōl > ol* is explained elsewhere (Phonology, § 9).

OMPTON.

1086 ⎰ Almuntone ⎱ D.B.
     ⎱ Almentone ⎰
1216–1307 Alemunton, Testa de N.
1278 ⎱ Almeton ⎰ H.R.
1316 ⎰   ⎱ F.A.
c. 1500 Elmeton, Inq. P.M. c. 1500.

[1] Cp. the German pl. n. *Kalter Herberg*, older *ze kalter herberge* (dative) in the Eifel district, Rhenish Prussia.

"The *tūn* or homestead of *Ealhmund.*" The development of the stressed vowel is similar to that in *Ollerton* (q.v.). It is noteworthy that in both pl. ns. the genitival *s* of the first element is absent from all the recorded forms.

The spelling *Elmeton* shows the influence of the independent word *elm-tree,* with which it was connected by popular etymology.

ORDSALL.

   1086 Ordeshale, D.B.
   1375 Ordesale, Index.
   1637 Ardsall, Map in Camden.

I take this name to be derived from O.E. *Ordrīces healh,* "the valley of *Ordric.*" A person of the latter name is said in D.B. to have held land in this locality. It is possible that he gave his name to the village (see *Gamston* near Retford).

The phonetic development can be easily explained. Of the two *r*'s the second one in the unstressed syllable was dropped (Phonology, § 14), whereas the fronted *c* would become assimilated to the following *s.*

Camden's spelling means nothing.

ORSTON.

Type I.

  1284 Orston ⎱
          ⎰ F.A.
  1428 Horston
 c. 1500 Horson, Inq. P.M. c. 1500.
  1637 Ouston, Map in Camden.

Type II.

  1086 Oschintone, D.B.
  1242 Orskinton, Inq. P.M. I.

From O.E. *Ordrīces tūn,* "the farmstead of *Ordric.*" Type II arose out of confusion with *Ossington* (q.v.). The *d* was lost between the two *r*'s at an early period ; the pers. n. itself occurs as *Orric* in O.E. The same name took a different line of change in *Ordsall* (q.v.). This variety of development may probably be accounted for by a different distribution of stress.

The initial *H*, which has no significance in this dialect, may

represent an attempt at connecting the name with *horse*.
Camden's form and that of 1500 are interesting in so far as they
may represent the contemporary pronunciation.

OSBERTON (under Scofton).

Type I.

1086 Osbernestune, D.B.
1428 Osberton, F.A.
1637 Osburton, Map in Camden.

Type II.

c. 1500 Esbarton, Inq. P.M. c. 1500.

O.E. $\bar{O}$*sbeornes tūn*, "the farmstead of $\bar{O}$*sbeorn*." This pers. n.
is of Scandinavian origin. Its prototype was $\bar{A}$*sbiorn*, but when
it was introduced into England, it became anglicised in form, the
ā (from *a* before *ns*) being changed to *ō* and *io* to *eo*. Type II
contains another Norse variant of the same pers. n. *Æsbiorn* (see
Björkman, p. 10). It is curious to meet Type II in so late a
document only; this seems to indicate that the scribes of the
15th century must have had access to old and reliable sources,
and that tradition in the spelling of pl. ns. was very strong.

The complete disappearance of the genitival *s* is a noteworthy feature of this name. The various forms assumed by *e*
before *r* are explained elsewhere (Phonology, § 8).

OSMONDTHORPE (under Edingley).

1086 Oswitorp[1], D.B.
1331 Osmundthorp, Index.
c. 1500 Ossonthorpe, Inq. P.M. c. 1500.

" The dwelling-place or hamlet of $\bar{O}$*smund*." This pers. n. is
found both in O.E. and O. Norse, in the latter language as
$\bar{A}$*smundr*, which, however, would readily assume the English
form (see preceding name). The nature of the second element
speaks in favour of Norse origin (see Björkman on *Asmund*,
p. 21).

The spelling of A.D. 1500 evidently represents the contemporary pronunciation, with *m* dropped in the beginning of
an unstressed syllable.

---

[1] Evidently a misreading; the scribe mistook *Osmutorp* for *Osuuitorp*.

OSSINGTON.

1086 Oschintone, D.B.
1173–4 Oskinton, P.R.
1216–72 Occington, Index.
1278 Oscington, H.R.
1327–77 Ossyngton, Non. Inq.

" The *tūn* or farmstead of *Ōsketin.*" This is the Scandinavian pers. n. *Āsketill,* of which numerous examples are found both in the original and the new domains of the Norsemen. The change from *ā* to *ō* is accounted for by the fact that the element *ās* was found as *ōs* in native Anglo-Saxon names (see preceding name). The substitution of the ending *in* for *ill* is explained by Dr Björkman (p. 17) as due to confusion of the two Latin suffixes *īnus* and *illus.* A similar transformation is noted by Professor Wyld, who finds the pers. n. *Rōskin* for *Rōsketill* in the Lancs. pl. n. Rossendale.

The phonetic changes are regular (see Phonology, §§ 1 ; 13). This name furnishes another instance of the transition of unstressed *in* to *ing* (Phonology, § 13).

OSWARDBECK (Wapentake).

Type I.

1086 $\left\{ \begin{array}{l} \text{Oswardebec Wap.} \\ \text{Wardebec Wap.} \end{array} \right\}$ D.B.

1153 Oswardebec, Index.
1189 Oswardesbech, P.R.

1278 $\left\{ \begin{array}{l} \text{Oswardebeck} \\ \text{Hoswordbec} \end{array} \right\}$ H.R.

Type II.

1216–1307 $\left\{ \begin{array}{l} \text{Oswaldebeck} \\ \text{Oswoldebeck} \end{array} \right\}$ Testa de N.

1444 Oswaldbeck, Index.

No attempt has been made to classify or discuss the variety of spellings found in the Inq. P.M. c. 1500. They afford an instructive example of what divergent results the united actions of phonetic tendencies and popular etymology may produce :

Oswaldbeck, Osberbeksoke, Hoswoldbekesoke, Oswalbeke,Oswardbek, Osilbeke, Ossonbek, Osbaldbekoop, Walbeksoken, Osylbeke.
This wapentake derives its name from a brook called *Ōsweardes bekk*, "the brook of *Ōsweard*" (Type I). As the second element is of Norse origin, O.N. *bekkr*, we may perhaps infer that the pers. n. was originally of the same origin. There exists a Scandinavian pers. n. *Āsvarðr*, of which traces are found in England (cp. Björkman, Index), and for which the native equivalent *Ōsweard* might have easily been substituted (see preceding names, and, on the subject of substitution in general, Björkman, pp. 197 sqq.).
In Type II the pers. n. *Ōsweald* is erroneously introduced.

OWTHORPE.
Type I.
1086 Ovetorp, D.B.
c. 1190 Hustorp, Woll. MSS. (queried by the editor).
1216–1307 Uvetorp, Testa de N.
1302 Outhorp, F.A.

Type II.
1284 $\left\{ \begin{array}{l} \text{Cupehoip} \\ \text{Cupthorp} \end{array} \right\}$ F.A.

"The thorpe, or dwelling-place of *Ūfi*, or *Ūvi*." The name is Norse in origin ; Dr Björkman quotes several instances of its occurrence in England. The *v* became vocalised after the *ū*. The initial *h* of the Woll. MSS. spelling means nothing (Phonology, § 19). I cannot explain Type II; it must have arisen out of a blunder of the scribe.

OXTON.
1086 $\left\{ \begin{array}{l} \text{Oxetune} \\ \text{Ostone} \end{array} \right\}$ D.B.
1278 Oxton, H.R.
1292 Oston, Index.
1302 Oxton $\left. \begin{array}{l} \\ \\ \\ \end{array} \right.$
1316 Hoxton $\left. \right\}$ F.A.
1346 Oxton

"The ox-enclosure." The name needs no further explanation.

PAPPLEWICK.

1086   Paplewic, D.B.
1189   Papewich, P.R.
1316   Papulwyk ⎫
1428   Papilwyk ⎭ F.A.

O.E. *papol wīc*, "the pebbly creek or bay." The village is situated on the eastern bank of the river Leen.

The second element, which is of Norse origin, is discussed under *Colwick*. *papol*(*-stān*) means "pebble" in O.E.

PERLETHORPE or PALETHORPE (Palethorpe, Hope).

1086   Torp, D.B.
1166–7   Peurelestorp, P.R.
1278   Pevereltorp, H.R.
         Peverltorp, Inq. P.M. II.
1316   Peverelthorp, F.A.
1637   Parlethorp, Map in Camden.

"The thorpe or village owned by the noble family of *Peverel*." The Peverels came over to England with the Conqueror, but apparently did not obtain land in Nottinghamshire until after the date of the Doomsday survey. Many places appearing simply as *Torp* in D.B. have later acquired a distinctive addition from the name of the then owner, usually a Norman nobleman, as *Thorpe Basset, Thorpe Mandeville* etc. It is rare, however, to find a Norman name prefixed in true Teutonic fashion as in the present name. A similar instance is found in *Cossardthorpe*, an ancient name for Hodsock which has not survived.

The curious development of the stressed vowel, *erl > arl > āl > eil* is a peculiar feature of the dialect (see Phonology, § 7).

The first modern spelling preserves an older type, whereas the second is phonetically correct.

PLUMPTREE.

1086   Pluntre, D.B.
1302   Plumtre, F.A.
1460 ⎧ Little Plumptree ⎫ Index.
       ⎩ Parvus Plomptre ⎭
1637   Plumbre, Map in Camden.

O.E. *æt plūm trēowe*, "at the plum-tree." In O.E. charters, trees are often referred to in connection with boundary marks and field names. The medial *p* represents the labial glide which developed between the *m* and the *t* (Phonology, § 16).

POULTER (river) [pautə].

RADCLIFFE-ON-TRENT [rætlif].

Type I.

$\left.\begin{array}{r}1086 \\ 1240\end{array}\right\}$ Radeclive $\left\{\begin{array}{l}\text{D.B.} \\ \text{Index.}\end{array}\right.$

1258 Radeklive, Inq. P.M. I.
1284 Radeclyve, F.A.

Type II.

1291 Radeclyf super Trent, Tax. Eccles.
1428 Radclif, F.A.
1637 Ratclyf, Map in Camden.

"The red cliff." Type I goes back to O.E. *æt þǣm rēadan clife*, whereas Type II, from which the modern spelling is derived, stands for the nominative *þæt rēade clif*. The village is situated on a lofty red cliff on the southern bank of the Trent. There is another place of the same meaning in this county (see *Ratcliffe*) and a *Radcliffe* in Lancashire. Corresponding German names also occur, as *Rothenfels* (Baden), O.H.G. *Rotenvels*, and *Rodestein*, called *zi themo roten stenni* in the older language (Förstemann, II).

The O.E. *ēa* was shortened at an early period before it had changed to M.E. *ē* (Phonology, § 1). In the combination *dkl*, the *k* caused the unvoicing of the preceding dental, after which it was dropped (Phonology, §§ 17; 12).

RADFORD (in Nottingham).

Type I.

1086 Redeford, D.B.

Type II.

c. 1240 Radeford, Bodl. Ch. and R.
1637 Radforde, Map in Camden.

O.E. *æt þǣm rēadan forde*, "at the red ford." The vowel of the first element appears long in D.B. (*e* in open syllable = [æ], Stolze, § 19), but was later shortened before the combination *df* (Type II). Retford in the north of the county has the same meaning; both places are situated on small water-courses just inside a stretch of Bunter sandstone. Especially in the neighbourhood of Radford the red colour of certain cliffs and of the river-bed is noticeable. It must have struck those coming out of the adjoining Keuper regions in particular as characteristic of the locality. There is a German pl. n. *Rotenförde* (Province of Saxony), older Low Germ. *Rodemfuordi* (Förstemann, II).

See *Retford*, and remark by Professor Wyld on Radcliffe (Lancs. Pl. Ns.).

RAGNALL.

Type I.

1086 Ragenehil, D.B.

1216–72 { Ragenil
          { Ragenhil } Index.

1329 Ragenhill

Type II.

1287 Reynilthorp, H.R.

It is impossible to explain this name satisfactorily. The second element may originally have been *þorp*, and the first a Scand. pers. n., either *Ragnald* or *Regnald*. The early spellings seem to show substitution of the Norse female name *Ragnhild* or *Regnhild*. (See Björkman, pp. 111, 112.) The confusion of the syllable *-all* (<*ald*) or *-hill* (<*hild*) with the frequent second theme O.E. *health* or *hyll* respectively may have led to the substitution of the latter and the dropping of what would then appear as a third element *-thorp*.

RAINWORTH (under Blidworth).

The second element is O.E. *weorþ, worþ*, "enclosed homestead, habitation." I cannot explain the first part, as there are no early forms. It may represent an old Celtic river-name (cp. the German river-names *Rhein, Regen*), or the first element of an O.E. pers. n. beginning with *Regin-*, such as *Regenbeald, Regenheard* etc.

RAMPTON.

$$1086 \begin{cases} \text{Rampestune} \\ \text{Rametone} \end{cases} \text{D.B.}$$

$$\left.\begin{matrix} 1302 \\ 1637 \end{matrix}\right\} \text{Rampton} \begin{cases} \text{F.A.} \\ \text{Map in Camden.} \end{cases}$$

" The *tūn* or homestead of *Hrafn.*" This Scandinavian pers. n. is found in various forms in English records, as *Rafn, Raven, Ram* etc. The latter type, with the articulations of *f* and *n* combined into *m*, is contained in the above pl. n. The development of a labial glide between *m* and *t* is a natural and regular process (Phonology, § 16).

RANBY.

$$1086 \begin{cases} \text{Ranesbi} \\ \text{Ranebi} \end{cases} \text{D.B.}$$

1316 Raneby, F.A.

" The *bȳ(r)* or dwelling of *Hrafn.*" The same pers. n. forms the first element of the preceding and following names.

RANSKILL.

       1086 Raveschell, D.B.

       1278 Ravenskelf, H.R.

       1704 Rawkild, Map.

" The well of *Hrafn.*" The second element is Scandinavian *kelde*, " a well," which is discussed under *Bothamsall* (q.v.). The pers. n. appears in the same form as in the preceding pl. n.

The H.R. spelling substitutes O.N. *skjalf*, older **skelf*, " a shelf, ledge, seat," for the original termination. This *skjalf* occurs in the Yorks. pl. n. *Ulleskelf*, " the ledge of *Ulfr* " (see Pl. Ns. of the W. Rid.).

aw in the spelling of 1704 seems to be the result of the vocalisation of *v* after *a* ; or is it a mistake ?

In the modern form, *v* has disappeared before *n* according to rule.

RATCLIFFE-UPON-SOAR.

Type I.

       1086 Radeclive, D.B.

       1189–99 Radeclivam super Soram, Index.

       1284 Radeclyve, F.A.

Type II.

1291 Radeclif super Soram, Tax. Eccles.
1637 Radclyff, Map in Camden.

Thoroton (I, 24) explains this name as meaning "*Red Hill or Bank.*"
See *Radcliffe* above.

REMPSTON.

1086 Repestone, D.B.
1155–65 Rempestuna(m), Nott'm Ch.
1302 Rempeston, F.A.
1327–77 Remeston, Non. Inq.
1637 Remston, Map in Camden.

"The *tūn* or homestead of *Reven.*" This pers. n. is a variant of *Raven*, from an original Scandinavian *Hrafn.* The development is similar to that of the same element in *Rampton* (q.v.), with this one exception, that in the present case the sign of the genitive has been preserved. Thoroton gives an alternative spelling *Rampeston* which shows that the two types of the pers. n. were interchangeable.

RETFORD (East and West) (Redfud, Hope).

Type I.

1086 Redforde, D.B.
1225   }
1227–77 } Retford { Bor. Rec.
                   { Non. Inq.
1278 Retteford, H.R.
1291 Retford, Tax. Eccles.
1316 Retteford, F.A.
1535 Redforth, Valor Eccles.
1704 Redford, Map.

Type II.

1155–65 }
1189   } Radeford, Nott'm Ch.

"The red ford." See *Radford*, which has the same meaning. The divergence of types is explained there. *d* has become *t* under the influence of the voiceless *f*. The late spellings containing *d* are due to attempts at etymological correctness. The suffix *-forth* of the Valor Eccles. is introduced from other pl. ns. which contain the Scandinavian *fjǫrðr* instead of the English *-ford*.

White's Directory (1853, p. 660) says that "the two Retfords were named after the ancient ford which crossed the Idle a little below the bridge which now unites them, and was called the *red ford* from its stratum of red clay being so frequently disturbed by the passage of cattle etc., as to tinge the water with its colour."

ROCKLEY (under Askham).

The second element seems to be O.E. *lēah*, "field, meadow." It is impossible to say what the first stands for; it may go back to O.E. *hrōc*, "rook (bird)," or M.E. *roc*, "rock," or it may contain the Scandinavian pers. n. *Hrōkr*, found as *Roc* in England.

ROLLESTON [roulstn].

$$1086 \begin{cases} \text{Roldestun} \\ \text{Rollestone} \end{cases} \text{D.B.}$$

1189 Rodeston, P.R.
1287 Rolliston, H.R.

$$\begin{matrix} 1302 & \text{Roldeston} \\ 1428 & \text{Rolleston} \end{matrix} \Big\} \text{F.A.}$$

1637 Rowlston, Map in Camden.

"The *tūn* or farmstead of *Rold*." This pers. n. is an abbreviation of the Scandinavian *Hrōaldr* (Björkman, p. 69). *ld* has become *ll*, after which change an *u*-glide developed between *o* and *ll* (Phonology, §§ 13; 9). The latter change is recognised by Camden but not in the modern spelling.

RUDDINGTON.

Type I.

$$1086 \begin{cases} \text{Roddintone} \\ \text{Rodintun} \end{cases} \text{D.B.}$$

c. 1190 Rudingtun, Woll. MSS.
1428 Rodyngton, F.A.
1637 Reddington, Map in Camden.

M.   8

Type II.

1227–77 Rotington, Non. Inq.
1261 Rotinton, Inq. P.M. I.
1291 Rotington, Tax. Eccles.
1302 Rotynton, F.A.

The first element is a patronymic in the genitive plural, derived from the O.E. pers. n. *Rudda.* The meaning of O.E. *Ruddinga tūn* is therefore "the homestead of the family of *Rudda*, the Ruddings" (Type I). A similar O.E. pers. n., *Ruta*, is contained in Type II; there must have been confusion between these two names. Camden evidently connects the name with *red*—another instance of his unrestrained etymological imaginativeness.

RUFFORD.

Type I.

(*a*) 1086 Rugforde, D.B.
    1155 Rūford, P.R.
(*b*) 1156 Rufford, Index.
    1278 Rafford, H.R.
    1291 Rufford, Tax. Eccles.
(*c*)   Rumford, Monasticon Anglicanum.

Type II.

1156 Rudford, P.R.
1275 Ruthford, H.R.

Type III.

1163 Rucford ⎫ P.R.
1198 Rocheforde ⎭
1637 Rucheforde, Map in Camden.

It is evident from the spellings under Type I *a*, that the first element is O.E. *rūh*, "rough." The meaning therefore is "the rough ford." The adjective may indicate either that the water was turbulent, or, more probably, that the ford was difficult to cross. There is a *Rufford* in Lancs., and a *Rufforth* in the West Riding.

The original *h* has become assimilated to the following *f*

(Type I *b*). Before the long *f*, the *ū* was shortened. Type I *c* represents the O.E. dative *æt rū(wu)m forde* (or *æt þǣm rū(wa)n forde*, with change of *nf* to *mf* through assimilation). The other types owe their origin to the peculiarities of Norman-French pronunciation. The sound of *h*, the guttural spirant, was unknown to the Normans, so they substituted *k* for it, as Englishmen will do at the present day with regard to German *ch* after back vowels. The *ch* of Type III stands for the sound of *k* as in many Norman records (Zachrisson, pp. 32 sqq.). By other Normans, the unfamiliar spirant was mistaken for the, to them, equally troublesome *ð*, which accounts for Type II.

The late appearance of Camden's form must be explained by assuming that he copied from an old source.

RUSHCLIFF (Wapentake).

> 1086 Riseclive, D.B.
> 1284 Riseclyve ⎫
> 1302 Ryseclive ⎬ F.A.
> 1428 Rysshclyve ⎭

This name needs no translation. Cliffs and mounds were favourite sites to hold meetings on (see *Bassetlaw*). The second element of the early spellings appears in the dative (see *Ratcliffe*).

The vowel of the O.E. *hrysce* is correctly represented by M.E. *i* in the above forms. The *u* of the modern spelling is due to the influence of the independent word, *rush*, introduced into the standard language from another dialect.

It may be mentioned here that the sound of *sh* (from O.E. *sc*) is very frequently represented by *s* in D.B. and other Norman records (Stolze, § 42).

SALTERFORD.

> 1086 Saltreford, D.B.

This name may stand for O.E. *sealtera ford*, "the ford of the salt-dealers." The manufacture and distribution of salt were of great importance in ancient times. Salt-springs, *salinae*, and "salt-streets" are frequently mentioned in mediæval records

8—2

(see Crawford Charters, p. 115). There is another Salterford in Worcestershire of apparently the same origin. Although the Notts. Salterford was situated in the very heart of the desolate forest, it is yet possible that one of the prehistoric cross-country tracks passed through the neighbourhood, and that this was frequented by salt-carriers.

Isaac Taylor (Engl. Village Names, § 5) derives this name from a hypothetical *sealh trēo ford*, "the ford by the sallow-tree." It is impossible to say which of the two explanations is the correct one.

SAUNDBY.

Type I.

$\left.\begin{array}{l} 1086 \\ 1189 \end{array}\right\}$ Sandebi $\left\{\begin{array}{l} \text{D.B.} \\ \text{P.R.} \end{array}\right.$

Type II.

$\left.\begin{array}{l} 1278 \\ 1346 \end{array}\right\}$ Saundebi $\left\{\begin{array}{l} \text{H.R.} \\ \text{F.A.} \end{array}\right.$

c. 1500 Saunby, Inq. P.M. c. 1500.

Type III.

1428 Saindeby, F.A.

Type I represents the original most faithfully. The first element may have been O.E. *sand*, "sand," so that the meaning would be "the sandy habitation."

I am, however, inclined to think that the Norse pers. n. *Sandi* is involved, of whose occurrence in England Dr Björkman (p. 116) gives one instance. The same name occurs as *Sanda* in O.E. and as *Sando* in O.H.G.

Norman influence is responsible for the change from *an* to *aun* in Type II which has survived (Phonology, § 11). *ain* from *an* represents a Central French sound-change; cp. French *pain* < **pane*, *laine* < *lana* etc. (Schwan-Behrens, Grammatik des Altfranzösischen⁶, 1903, § 53, 1 a).

SAXONDALE or SAXENDALE.

Type I.

1086 Saxeden, D.B.

Type II.

1284 Saymdall, (?) F.A.
1291 Saxindal, Tax. Eccles.
1302 } Saxendale { F.A.
1472 } { Index.
c. 1500 { Saxondale } Inq. P.M. c. 1500.
       { Saxendall }
1637 Saxindale, Map in Camden.

Type III.

c. 1500 Saxbye, Inq. P.M. c. 1500.

It is impossible to say what the O.E. prototype of this name was. Whereas the second element of Type I is of O.E. origin, Type II, which survived, contains its Norse equivalent, both *denu* and *dalr* meaning the same thing, viz. "valley." The third type contains another Norse word, *bȳr*, "habitation, village."

The first element, too, is of doubtful origin. There are three possibilities; it may stand for: (1) the genitive of O.E. *Seaxe*, "Saxons," which was *Seaxna* (Sievers, § 276, anm. 3 *a*); (2) the genitive of the O.E. pers. n. *Seaxa*, or (3) of the Scandinavian pers. n. *Saxi.*

Interchange of the suffixes *den* and *dale* is also found in the early forms of the Lancs. pl. ns. Skelmerdale and Ainsdale.

SCAFTWORTH.

1086 Scafteorde, D.B.
1227–77 Skaftwurth, Non. Inq.
1278 Skastewurh, H.R.
c. 1500 { Skastworth }
       { Scastworth } Inq. P.M. c. 1500.
       { Scarworth (!) }

"The *weorþ* or farmstead of *Skafti.*" The pers. n. is of Norse origin as is proved by the initial *sk*. If the O.E. corresponding form *Sceaft* (Onomasticon) were contained in this name, the initial *Sc* would be pronounced *sh*. It is, however, quite possible that the O.E. form was the original one for which the Scandinavian equivalent was substituted later on.

The name *Skapti* is not recorded by Dr Björkman as found in England. It occurs frequently in Iceland (cp. Landn. Bōk, 79) and also in Norway (Rygh, G. Personnavne, p. 219).

In the above spellings, *st* may be due to a misreading, the original *f* being taken for the long *s*. Are the modern editors responsible for this blunder?

SCARLE.

1086 { Scornelei, D.B. (Victoria County History).
      { Scorvelei, D.B., as transcribed by Thoroton.
1227–77 Southscharle, Non. Inq.
1316 Scarle, F.A.

The second element is O.E. *lēah*, "field." The first seems to stand for a pers. n. If the D.B. spelling as read by Thoroton is correct, it represents the O.E. pers. n. *Sceorf*. The later spellings and the modern form cannot, however, be directly descended from an O.E. *Sceorfes lēah*. The initial *sk* clearly points to Scandinavian influence (cp. preceding name). One is, therefore, compelled to assume that the Scandinavian pers. n. *Skarf* (Björkman, p. 122) was substituted, which also accounts for the change in vowel.

v was lost between *r* and *l* as the result of a general tendency (Phonology, § 12).

SCARRINGTON.

1086 Scarintone, D.B.
1242 Scherinton, Inq. P.M. I.
1637 Sharinton, Map in Camden.

The first element seems to be the genitive plural of some patronymic, of Scandinavian origin, as is clearly shown by the initial *sk*. The recorded spellings are insufficient to form an opinion as to the name involved: it may have been *Skarf* (Björkman, p. 122) or *Skarði* (Rygh, G. Personnavne, p. 220).

SCOFTON.

1086 Scotebi, D.B.

The initial *sk* proves that the first element is a Scandinavian pers. n., most probably *Skopti*, recorded by Rygh (G. Personnavne, p. 225). The meaning therefore is " *Skopti's* farmstead."

SCREVETON [skrītn].

Type I.

1086 { Screvetone
        Screvintone } D.B.
        Escreventone }
1302 Screveton, F.A.

Type II.

c. 1500 Screton, Inq. P.M. c. 1500.
1535 Scretton, Valor Eccles.

Type III.

1284 Scrouton, F.A.
1637 Skiwton (?), Map in Camden.

The first element must be of Norse origin as is demonstrated
by the initial *sk* (cp. *Scaftworth*). It probably represents the
genitive of a pers. n. following the weak declension. What this
name was, I am unable to say. Type III cannot easily be
reconciled with the rest.

SCROOBY.

1086 Scrobi, D.B.
1227-77 }
  1278 } Scroby { Non. Inq.
                   { H.R.

c. 1500 { Scrobia }
          Scrobi   }
          Scrowby } Inq. P.M. c. 1500.
          Scroby   }

The first element may contain the O.N. pers. n. *Skorri*,
which Dr Björkman finds in the pl. ns. *Skorreby*, *Scorby*, *Skorton*
(p. 124). Metathesis of *r* frequently occurs in this dialect
(Phonology, § 15). The meaning is, therefore, "*Skorri's* farm,
or hamlet."

The modern spelling is misleading. It perpetuates a M.E.
habit of writing *oo* both for O.E. *ō*, and *ǭ* from *o* in open syllable.
One would expect the pronunciation to be [skroubi].

SELSTON.

1086 Salestune, D.B.

$\left.\begin{array}{l}1284 \\ 1291\end{array}\right\}$ Seliston $\left\{\begin{array}{l}\text{F.A.} \\ \text{Tax. Eccles.}\end{array}\right.$

1316 Selleston, F.A.

The first element seems to be a pers. n. in the genitive case; perhaps O.E. *Selua*, or Scand. *Seli(r)*?

SERLBY (Sarl-by, Hope).

1086 Serlebi, D.B.

1302 Serleby, F.A.

1637 Surlbye, Map in Camden.

" The *bȳ(r)* or farmstead of *Serlo*." The suffix as well as the pers. n. are from O. Norse. Dr Björkman (p. 117) refers to this pl. n. as containing the above Scandinavian pers. n.

For the development of *er* into *ur, ar*, see Phonology, §§ 7 ; 8.

SHELFORD.

1086 Scelforde, D.B.

$\left.\begin{array}{l}1278 \\ 1284\end{array}\right\}$ Schelford $\left\{\begin{array}{l}\text{H.R.} \\ \text{F.A.}\end{array}\right.$

The name *Sceldfor* is found on a coin struck about the year 890, and it is conjectured that this stands for the above pl. n. (Onomasticon, s.v. *Sihtric comes*). If this is correct, the etymology at once becomes clear. There existed in O.E. the adjective *sċeald*, W.S. *sċield*, "shallow," which appears in M.E. as *shoal*, from the Anglian type (see Napier and Stevenson, Transactions of the Philol. Soc. 1895–8, 532; Ekwall, Beibl. z. Anglia, XX, 209; Schlutter, Engl. Stud. 43, 318). I assume that by the side of the West Germanic adjective *skalða-* there existed also a form *skalðja-*, which would produce Anglian *sċelde* (Bülbring, § 175 anm.). The relation between the two types would be the same as that between O.E. *smolt* and *smylte*, " quiet," *strong* and *strenge*, " strong," etc. (Sievers, § 299, anm. 1).

I take, therefore, the above name to stand for O.E. (Anglian) *æt þǣm *sceldan forde*, "at the shallow ford." After the disappearance of the adjectival ending, *d* would drop in the combination *ldf*.

The name *Scealdan ford*, "at the shallow ford," occurs in an O.E. charter (Cart. Sax. 758; 802). Searle is certainly wrong in explaining the first element as a pers. n. The unmutated form of the adjective, from *skalŏa-, seems to be contained in the modern pl. n. *Shalford* in Surrey.

The corresponding form of the mutated adjective seems to be contained in the Hessian name *Schöllenbach*.

The following description of the neighbourhood is calculated to support the proposed etymology : "Shelford...is a pleasant village on a gentle eminence, which in very great floods is sometimes completely surrounded by the Trent water...though it is distant half a mile from the regular channel of the river..." (White, Directory, 1853, p. 455). The Trent has evidently changed its course in this locality.

SHELTON.

$$1086 \begin{cases} \text{Sceltun} \\ \text{Sceltone} \end{cases} \text{D.B.}$$

1302 Schelton, F.A.

1637 Shilton, Map in Camden.

The village is situated on a ridge overlooking the river Smite. I am, therefore, inclined to connect the first element with O.E. (W. Sax.) *scylfe*, Anglian *scelfe*, "shelf, ledge." The meaning, therefore, is "the *tūn* or homestead on the ridge."

SHERWOOD FOREST.

1189 Schirewude, P.R.

1272 Syrewde forest, Inq. P.M. I.

1278 Shirwod, H.R.

1393 Shyrewode, Index.

$$1637 \begin{cases} \text{Shirewood, Camden, p. 547.} \\ \text{Sherewood, Map in Camden.} \end{cases}$$

Camden says that "some expound [this name] by these Latin names *Limpida Sylva*, that is, *A Shire* or *Cleere wood*; others *Praeclara Sylva*, in the same sense and signification" (p. 550). It seems highly improbable, however, that it has anything to do with the O.E. adjective *scīr*, "bright, pure." When the name was first given, the mediæval mind had not yet awakened to a sense of the beauties of the primeval forest. On the contrary, large and dense woods such as this one inspired superstitious fears ; they were regarded as inimical to civilisation, the seat of man's worst enemies: "vasta solitudo,...saltus ferarum et cubile draconum" are the terms used by a German monastic chronicler[1] in reference to a wood near Berchtesgaden. Others connect the first element with modern *shire*. In the earliest records, Sherwood is often spoken of as the "forest of Nottingham" (Victoria County Hist. I, 365), which would seem to support the derivation from *shire-wood*, "the wood belonging to, or forming part of, the county."

But this is not wholly satisfactory either. I venture to suggest that the word *scīr-* is used here in the same sense as in *Shireoaks* (q.v.), and Shire Dyke, a little stream forming part of the boundary between the counties of Nottingham and Lincoln. Its meaning is "boundary, division." Jellinghaus (p. 316) connects the word with modern Westphalian *Schier*[2], of the same meaning, which enters into numerous Low German field names, such as *Schiereneiken*, "Shire-oaks," *Schierenböken*, "-beeches," *Schierholz*, "-holt, or wood." This last name is repeated in the O.E. *scirholt* quoted from a charter in Jellinghaus' article[3]. There exists also a *Shirland* in Derbyshire, which goes back to older *Scirlund* (Inq. P.M. 56 He. III), *lund* being the Scandinavian word for "wood." It may be noted here that O.E. *scīr-*, "boundary," is not connected with O.E. *scieran, sceran*, "to cut, shear"; Prof. Skeat in his Etymological Dictionary (1910, s.v. *shire*) remarks that its root is unknown.

If this explanation is adopted, the meaning would be "boundary forest." This seems a most appropriate name, seeing that Sherwood Forest stretches along the boundary between Nottinghamshire and Derbyshire, and that part of its ancient bounds, as laid down in the perambulations, actually coincides with the modern line dividing the two counties. Moreover, we learn from Tacitus (Germania, XL, 1; Annales, I, 61) that dense, impenetrable forests were looked upon as the safest boundaries

---

[1] See Gertrud Stockmeyer, Das Naturgefühl in Deutschland im 10. u. 11. Jahrhundert. Leipzig u. Berlin, 1910, p. 8.

[2] O. Saxon *ī* remains unchanged in Low German; see Herm. Teuchert, Laut und Flexionslehre der Neumärkischen Mundart, § 55, Zeitschr. f. deutsche Mundarten, 1907-8.

[3] Cp. also the expression *andlang scīre on hweðels heal*, Cod. Dipl. 5, 358, 15.

by the Germans. Sherwood Forest certainly was of that character. Boundaries separating peoples and tribes so frequently coincided with forests that primitive Germanic *markō-, Gothic *marka*, "boundary," actually changed its meaning in the Scandinavian languages, which use the word *mark* in the sense of "wood." In the O.H.G. fragment of a geographical didactic poem known as "Merigarto," a passage is found illustrating this function of large tracts of wood-covered land, which I cannot refrain from quoting. It runs as follows:

"michili perga skinun duo an der erda.
die sint vilo hôh, *habant manigin dichin lôh.*
daz mag man wunteren daz dâr ie ieman durh chuam.
dâmit sint dei rîche giteilit ungelîhe[1]."

The phonetic development offers no difficulties. The vowel *ī* was shortened before the combination *rw*; later on, *ir* and *er* were levelled under one sound, a change which is reflected in the modern spelling. For a similar development cp. *sheriff*, from O.E. *scīr-gerēfa*.

SHIREOAKS (Shireaks, Hope).

1216–1307 { Scirop } Testa de N.
              { Chirbrok }
1272–1307 Shiroaks, Index.
1458 Schyroks, Bodl. Ch. and R.
1535 Sirokks, Valor Eccles.
1637 The Shireokes, Map in Camden.

The name in Camden's Map does not seem to apply to a village but rather to a district. The "Shire Oaks" probably were a number of trees or a small copse near the boundary of the county. Legend speaks of *one* tree only as having given its name to the locality[2]. All the spellings, however, are in the plural. John Evelyn in "Sylva" (1664) has an interesting note on this supposed tree which, however, he knew by hearsay only. He writes: "*Shireoak* is a tree...which drops into 3 shires, viz. York, Nottingham and Derby."

---

[1] "Large mountains appear there on the earth. These are very high, *they have many a dense forest*. One may well wonder that man ever penetrated them. *By these the kingdoms are divided unequally.*"

[2] See J. Stacye in White's "Dukery Records," pp. 70 sqq.

The pl. n. *Skyrack* in Yorks. has a similar meaning, but is entirely Scandinavian in form, as appears still more clearly in the D.B. spelling which is *Schyrayk*. The above spellings of Testa de N. are blunders due to false analogy.

SIBTHORPE.

1086 Sibetorp, D.B.
1199–1216 Sibbetorp, Index.
c. 1200 Sibetorp, Woll. MSS.
1302 Sibbthorp, F.A.

"The habitation or village of *Sibba* or *Sibbi*." The pers. n. may be of O.E. or Scandinavian origin. The nature of the second element speaks in favour of the second alternative. *Sibbi* is, however, not mentioned by Dr Björkman as a Norse name found in England.

SKEGBY.

1086 $\left\{\begin{array}{l}\text{Schegebi}\\\text{Schachebi}\end{array}\right\}$ D.B.

$\left.\begin{array}{l}1302\\1316\end{array}\right\}$ Skegby, F.A.

"The *bȳ(r)* or dwelling of *Skeggi*." The pers. n. as well as the second element is of Norse origin. The meaning of the former is "the bearded one."

SMITE (river).

1535–43 $\left\{\begin{array}{l}\text{(a praty Broke or Ryveret}\\\text{caullit) Myte}\end{array}\right\}$ Leland, I, 106.
c. 1613 Snite, Drayton's Polyolbion, 26, 32.
1637 Snite, Map in Camden.

In O.E. we find the word *smīta*, "a foul, miry place." See Cod. Dipl. III, 166; 2–20; V, 105; 13–36. This is connected with O.E. *smītan*, "to daub, smear, pollute." If this word or some other derived from the same root is contained in the above name, the sense would be "dirty, miry water, or stream."

The omission of initial *s* in Leland's form is remarkable. Can he have copied it from a Norman-French document? Cp. the loss of *s* in *Nottingham*. The change of *m* to *n* after *s* is due to assimilation.

SNEINTON [snentn].

Type I.

(a) 1168–9 ⎫ Snotinton ⎰ P.R.
     1205 ⎭   ⎱ Cal. Rot. Chart.
(b) 1215 Snoditon, Cal. Rot. Chart.
(c) 1278 ⎫   ⎰ H.R.
     1316 ⎬ Sneynton ⎨ F.A.
     1571 ⎭   ⎱ Index.

Type II.

1086 Notintone, D.B.

O.E. *Snotinga tūn*, "the homestead of the *Snotings.*" The same family that settled at *Nottingham* (q.v.) seems to have founded a village here. Type II represents the Norman pronunciation with the initial *s* dropped as in *Nottingham.* As, however, the Norman element was not so overwhelmingly strong in this place as in the neighbouring fortified town, the native form prevailed.

The phonetic development of this name presents several interesting features. *ng* [ŋ] became assimilated to the following *t*; the first *t* apparently was voiced under the influence of the surrounding vowels (Type I *b*). After that change it disappeared, so that the vowels of the first and second syllables, *o* and *i*, collided and formed a diphthong. The diphthong *oi* being unfamiliar to the English-speaking population—M.E. *oi* is of French origin—it was soon replaced by *ei* which occurred in their language as the descendant of O.E. *eg, æg*, and Scandinavian *ei.* This diphthong which is preserved in the modern spelling was monophthongised to *ē* probably in the 15th century (Horn, §114). This latter sound was shortened before the combination *nt* in pronunciation, the result being [e] as in *says* [sez], *said* [sed] etc.

The Norman form of D.B. seems to have been preserved in the name of *Notintone Place* in Sneinton.

SOAR (river).

1253 Sor, Cal. Rot. Chart.

This may be a Celtic river-name; Mr McClure (p. 264, note 2) proposes to connect it with the *Sarua* of the Ravenna

Geographer. A name *Sordic* (-ditch) is mentioned in the index of Kemble's Cod. Dipl., and there exists a *Sorbrook* in Oxfordshire. On the continent, one finds several river-names compounded with *Sor-*, as O.H.G. *Soraha, Sorna* (Förstemann, II).

SOUTHWELL [saðə] (Suthull, Hope).

Type I.

958   at Suðwellan, Cart. Sax. 1029.
1086   Sudwelle, D.B.
1130   Sudwell ⎱ Index.
1331   Suthewell ⎰
1637   { Southwell ⎱ Camden, p. 549, and Map.
        { Suthwell ⎰

Type II.

1278 ⎤   ⎧ H. R.
1291 ⎬ Suwell ⎨ Tax. Eccles.
1323 ⎦   ⎩ Bor. Rec.

The etymology is obvious. " The modern name of the town is supposed to have arisen from a spring or well on the south side of the church; now called Lady Well and Holy Well, a noted spring situated on the right of the cloisters" (White, Directory, 1853, p. 509). It is equally, if not more probable that the name was given in contradistinction to *Norwell*, "the north well," some seven or eight miles to the north-east[1]. The shortening of the vowel (O.E. *ū* in *sūð*), the loss of initial *w* and final *l* are explained elsewhere (Phonology, §§ 1; 18; 21). Type II, which has not survived, is an interesting example of the loss of *th* [þ] before *w*, which is also encountered in the modern pronunciation of *southwester* [sauwestə] (Horn, § 201).

SOWLKHOLME or SOOKHOLME [saːkm].

1189   Sulcholm, P.R.
1272–1307   Sulholm, Inq. P.M. II.
c. 1500   Solcome, Inq. P.M. c. 1500.
1553   Suckholm, Index.
1637   Sowcam, Map in Camden.

---

[1] " The church of Southwell had possessed a manor at Norwell before the Norman Conquest " (Victoria County Hist. II, 153).

O.E. *sulh cumb*, " miry, wet valley." Cp. *sulig cumb*, Cart. Sax. 589. The second element, O.E. *cumb*, "a deep, hollow valley," was originally borrowed from some Celtic language. The first part, which does not seem to occur independently in O.E. literature, is identical with O.H.G *sulag*, "miry pool, volutabrum," found by Förstemann (II) in numerous pl. ns. such as *Solach* near Tegernsee from O.H.G. *Suligiloch*[1].

The spellings in -*holm* are due to confusion with the pl. n. element *holme*. Camden's form shows that, after the loss of the *h*, the *l* had become vocalised ; *ow* stands for M.E. $\bar{u}$ ($< u + u\,(l)$); this $\bar{u}$ was shortened as in the Index spelling of 1553 and the modern pronunciation. This interpretation of the pl. n. is borne out by a description of the locality. "A quarter of a mile S.W. of the village, is an excellent spring of water, where formerly was a bath ; from it a small stream runs through the village, and joins the Meden from Pleasley" (White, Directory, 1853, p. 653).

SPALFORD.

1086 Spaldesforde, D.B.
1302 Spaldeford, F.A.

If the first element is a pers. n. as seems to be indicated by the presence of the genitival *s*, it would be an O.E. *Spald(a)*, which is not recorded in the Onomasticon, but assumed to have existed by Prof. Skeat on the evidence of its occurrence in pl. ns. (Pl. Ns. of Huntingdon, p. 352). The meaning would be " Spalda's ford." This pers. name might be identical with the early Germanic *Spatalus* quoted by Werle (p. 54), which looks like a diminutive in -*al* derived from the ancestor of O.H.G. *Spatto* (Förstemann). *tl* becomes *ld* in certain O.E. dialects (cp. *botl—bold*, Sievers, § 196, 2).

There is another possibility. The *s* in D.B. may very well be spurious, cases of the insertion of an inorganic *s* by the compilers of that survey being very numerous (Zachrisson,

---

[1] In their edition of the Crawford Charters, Messrs Napier and Stevenson (p. 47) explain *Sulhford* as " a ford approached on one or both sides by a sunk road or gully." I do not agree with their interpretation, but prefer to connect this name too with the O.H.G. word (see also Jellinghaus, p. 317).

pp. 118 sqq.) If that is so, the first element might represent O.E. *spātl*, *spāld*, which two forms stand in the same relation as *botl*, *bold* quoted above. *spāld* means "saliva," but as there exists a verb *spǣtlan*, "to spit foam," we may infer that *spāld* could be used in the sense of "foam" as well. The name "foam(y) ford" seems a very natural one.

The shortening of the vowel *ā* and the loss of *d* are the results of natural tendencies (Phonology, §§ 1; 12).

STANFORD-ON-SOAR.

$$\left.\begin{array}{l} 1086 \\ 1302 \end{array}\right\} \text{Stanford} \left\{\begin{array}{l} \text{D.B.} \\ \text{F.A.} \end{array}\right.$$

"The stone ford," O.E. *stān ford*. The name applies either to the condition of the river-bed, or to stepping-stones, by means of which the ford was crossed. There are fifteen Stanfords, Stamfords or Stainforths in England, the last-named being Scandinavian in both elements. Isaac Taylor (Engl. Village Names, § 5) says that they were so named because they were "paved with stones." The name *Steinfurt* is found in Germany.

O.E. *ā* was shortened before the combination *nf* previous to becoming rounded in early M.E. (Phonology, § 1 *a*)

STANTON-ON-THE-WOLDS.

1086 Stantun, D.B.

$$\left.\begin{array}{l} 1189 \\ 1222 \end{array}\right\} \text{Stanton} \left\{\begin{array}{l} \text{P.R.} \\ \text{Bor. Rec.} \end{array}\right.$$

$$\left.\begin{array}{l} 1222 \ \text{Estanton} \\ 1240 \ \text{Stanton-super-Wold} \end{array}\right\} \text{Bor. Rec.}$$

1637 Stannton, Map in Camden.

O.E. *stan tūn*, "the stony homestead, the village on the stony land." The country round about Stanton is extremely bleak, and the land " of a sandy wet quality " (White, Directory, 1853, p. 404).

For the development of O.E. *ā* see *Stanford*. *Estanton* is a Norman-French form, with *e* prefixed to the initial combination *st* in conformity with a general tendency prevalent among French speakers (Zachrisson, pp. 55 sq.).

STAPLEFORD [stæplfəd].

1086 Stapleford, D.B.

$\left.\begin{array}{c} 1254 \\ 1284 \end{array}\right\}$ Stapilford $\left\{\begin{array}{l} \text{Index.} \\ \text{F.A.} \end{array}\right.$

$\left.\begin{array}{c} 1348 \\ 1356 \end{array}\right\}$ Stapulford, Index.

"The ford by or leading to the pillar." The name is derived from the stone cross which still stands near the church, and has been declared to go back to Anglo-Saxon times to a date not later than the ninth century (Guilford, p. 187). O.E. *stapol* means "a pillar, boundary mark." In old German law, the word *stapol* had a special sense: it denoted the pillar near which the courts assembled and where judgment was given. This signification may also have been possessed by the *stapol* from which this pl. n. is derived (see Grimm, Rechtsaltertümer, p. 804). Förstemann (II, s.v. *Stapf*) quotes the following passage: "ad regis staplum, vel ad eum locum, ubi mallus est." Ducange (Gloss. Mediae et Infimae Latinitatis) explains "mallus" as meaning "Publicus conventus, in quo majores causae disceptabantur, judiciaque majoris momenti exercebantur a Comitibus, Missis dominicis, aliisque Judicibus."

Prof. Skeat denies that the word *stapol* could be applied to a stone pillar. He says: "A.S. *stapol* simply means a wooden post or pole; and Staple-ford merely means that such a post marked the position of the ford. Where is the evidence that it ever meant a sculptured pillar? I take it to be all a fantastic dream..." (Notes and Queries, 11, S. 11, 1910).

Prof. Skeat's view is supported by the fact that there exist many other *Staplefords* in other counties where there are no stone crosses. Both interpretations may be right, so that, until further evidence is adduced, the reader must choose between the two possibilities as the fancy takes him.

STAUNTON.

Type I.

$\left.\begin{array}{c} 1086 \\ 1637 \end{array}\right\}$ Stanton $\left\{\begin{array}{l} \text{D.B.} \\ \text{Map in Camden.} \end{array}\right.$

M.

Type II.

$\left.\begin{array}{l}1216-1307 \\ 1302\end{array}\right\}$ Staunton $\left\{\begin{array}{l}\text{Testa de N.} \\ \text{F.A.}\end{array}\right.$

This name is identical in origin and meaning with *Stanton*, above. Type II, which persists in the current modern form, represents the Norman-French pronunciation of this name, with *aun* for *an* (see *Saundby*).

STAYTHORPE.

Type I.

c. 1175 Stiresthorp, Woll. MSS.

Type II.

(*a*)   1086 Startorp, D.B.

$\left.\begin{array}{l}1278 \\ 1302\end{array}\right\}$ Starthorp $\left\{\begin{array}{l}\text{H.R.} \\ \text{F.A.}\end{array}\right.$

$\left.\begin{array}{l}1302 \\ 1346\end{array}\right\}$ Sternethorp $\left.\begin{array}{l}\\ \\ \end{array}\right\}$ F.A.
1346 Starthorp

1412 Sternethorp, Index.

c. 1500 $\left\{\begin{array}{l}\text{Sterthorp} \\ \text{Starethorp}\end{array}\right\}$ Inq. P.M. c. 1500.

1535 Stertherop, Valor Eccles.

(*b*)   c. 1500 Stathorpe, Inq. P.M. c. 1500.

"The habitation or village of '*Styr.*'" This Scandinavian pers. n. is most faithfully preserved in Type I; it is a nick-name in origin, meaning "strife, battle." English forms of the same pers. n. are *Ster*, *Sterre* enumerated by Dr Björkman (p. 132). These latter have been substituted for the original *Styr* (*Stir*) in Type II. The *n* which occurs in several spellings is the ending of the weak genitive. *ar* is regularly developed from *ær*; *ar* becomes *ā* before open consonants (Type II *b*, Phonology, § 7). The latter form survives in the modern spelling, *ay* standing for M.E. *ā*, now pronounced [ei].

STOCKWITH (West).

No early forms. For a discussion of the first element see the name below. The second theme may be either Scand. *vað*, "ford," or *viðr*, "tree, wood." *Stocc wað* might mean "the ford indicated by a "stock" or pole; compare *Langwith*, and *Stapleford*.

STOKE BARDOLPH.

1086 Stoches, D.B.
1302 Stok, F.A.
c. 1500 Stokkerdolffe, Inq. P.M. c. 1500.

The original meaning of O.E. *stocc* is "stock, log, stump of a tree." Isaac Taylor (Engl. Village Names) says concerning the numerous places called Stoke that they derive their name from their position "near the stump of a tree in a half-cleared forest." Others believe that *stocc* was used to denote a "fenced-in place," i.e. an enclosure secured by means of "stocks" or wooden palings (Alexander, Pl. N. Oxfordshire, p. 196). It is also conjectured that *stocc* had the meaning not only of "log," but also of a collection of such, i.e. "a log-cabin" or "block-house." But why should the vowel in all the modern names be long, when the O.E. prototype contained a long or double consonant *cc*?

Bardolph is the name of a former owner added in order to distinguish the place from the other Stokes. The Inq. spelling records a curiously corrupted pronunciation.

STOKE (EAST).

1086 { Stoches / Estoches } D.B.

1302 } Stok { F.A.
1273–1307 } { Index.
1586 East Stoake, Index.

See preceding name. The initial *e* of the second D.B. form is not a remnant of a prefixed distinctive *east* but owes its origin to a Norman-French habit of speech; cp. French *esprit* from Latin *spiritus* etc.

STOKEHAM.

1302 Stocum, F.A.
1412 Stokum, Index.

O.E. *æt stoccum*, "at the tree stumps," or, "at the log-cabins"? The dative plural of O.E. *stocc*, of uncertain meaning. See *Stoke Bardolph*.

STRAGGLETHORPE (under Cotgrave).

There are no early forms. Can the first element be a corruption of a Scandinavian pers. n. *Strangulfr?

STRELLEY.

Type I.

$$1086 \left\{ \begin{array}{l} \text{Straleia} \\ \text{Straelie} \end{array} \right\} \text{D.B.}$$

$$\begin{array}{l} 1166\text{-}7 \quad \text{Stratlega} \\ 1189 \quad \text{Stradlega} \end{array} \right\} \text{P.R.}$$

Type II.

(*a*) 1189 Stretlee, Nottm. Ch.

$$\left. \begin{array}{l} 1216\text{-}1307 \\ 1275 \end{array} \right\} \text{Stretleg} \left\{ \begin{array}{l} \text{Testa de N.} \\ \text{F.A.} \end{array} \right.$$

$$(b) \quad \left. \begin{array}{l} 1291 \\ 1302 \\ 1428 \end{array} \right\} \text{Stredley} \left\{ \begin{array}{l} \text{Tax. Eccles.} \\ \text{F.A.} \\ \text{F.A.} \end{array} \right.$$

$$(c) \quad \left. \begin{array}{l} 1284 \\ \text{c. } 1500 \end{array} \right\} \text{Strelley} \left\{ \begin{array}{l} \text{F.A.} \\ \text{Inq. P.M. c. } 1500. \end{array} \right.$$

Type III.

$$\text{c. } 1500 \left\{ \begin{array}{l} \text{Stertley} \\ \text{Sterley} \end{array} \right\} \text{Inq. P.M. c. } 1500.$$

O.E. (Mercian) *on strēt lēʒe,* "in the field by the street." The O.E. (W. Sax.) *strǣt* originally comes from the Latin *strāta via* and is usually employed of a Roman road. Such a one must have run past Strelley.

Type I comes from the W. Saxon variant of the O.E. word, *a* being the result of the shortening of older *ǣ*. The other types contain *e* derived by the same process from the native Mercian *ē*. The development of the *t* may be traced in its various stages through the early spellings. It is preserved intact in Type II *a*; then it becomes voiced under the influence of the surrounding sounds (Type II *b*), and is finally assimilated by the following *l* (Type II *c*, Phonology, § 13).

Type III, which perished, shows metathesis of *r*.

The pl. n. *Streatley* occurs in Bedfordshire.

STURTON-IN-THE-CLAY or STURTON-LE-STEEPLE.

Type I.

1166–7 Strotton, P.R.

c. 1200 Strattone, Cal. Rot. Chart.

Type II.

1086 Estretone, D.B.

1216–1307 Strecton, Testa de N.

1278		H.R.
1291	Stretton	Tax. Eccles.
c. 1300		Index.
1302		F.A.

1375 Neyerstretton	
1383 Ouerstretton	Index.
1384 Stretton en le Clay	
1425 Stretton in the Clay	

c. 1500 Stretton, Inq. P.M. c. 1500.

Type III.

c. 1500 Stirton, Inq. P.M. c. 1500.

O.E. (Mercian) *strēt tūn,* "the homestead by the street." The Roman road to which the name refers is the one leading from Lincoln to Doncaster. Type I is to be explained in the same way as Type I of *Strelley* (q.v.). The P.R. spelling of *o* for *a* is a scribal error; so is the *c* which stands for *t* in Testa de N. The initial *e* of D.B. is Norman-French in origin. The *y* in the Index form of 1375 is a M.E. spelling for *þ, th;* *u* in the form of 1383 of course stands for *v.*

Type III which prevailed shows metathesis of *r* (Phonology, § 15). *er, ir, ur* all represent one and the same modern sound (Phonology, § 8).

The distinctive addition—Norman-French *en le Clay,* English *in the Clay*—indicates the nature of the soil. Sturton is in the North Clay Division of Bassetlaw Hundred.

"Sturton-le-Steeple owes the latter part of its name to the far-seen array of twelve pinnacles with which the builders thought fit to surround the parapet" of the church tower (A. Hamilton Thompson, in "Memorials of Old Nottingham-shire," p. 52).

STYRRUP or STYRUP.

Type I.

1086 Estirape, D.B.

1278 Stirap, H.R.

c. 1300 Styrap, Index.

1302 }
1348 } Stirap { F.A.
     { Index.

Type II.

1414 Sterap, Index.

c. 1500 Sterop, Inq. P.M. c. 1500.

"The valley of *Styr*." This O.N. pers. n. has been discussed under *Staythorpe* (q.v.). An English variant of the same name, *Ster*, accounts for Type II. The second element represents O.E. **hop*, which is found as an independent word in M.E. only, meaning "valley, hollow among hills."

The initial *e* of D.B. is Norman-French in origin.

SUTTON (near Granby).

1179 Suttuna, Index.

1284 Sotton, F.A.

SUTTON-IN-ASHFIELD.

1086 Sutone, D.B.

1316 { Sutton super Asshefeld } F.A.
  { Sutton super Essefeld }

1535 Sutton super Lownde, Valor Eccles.

SUTTON BONINGTON.

1338 Sutton super Soram } Index.
1395 Sutton super Sore }

SUTTON-UPON-TRENT.

1412 Sutton, Index.

O.E. *sūþ tūn*, "the southern farmstead." It is, of course, impossible to say with regard to which place or object this name was originally given. It is one of the commonest pl. ns. in England.

The shortening of O.E. *ū* before *tt*, the result of assimilation of *þ* by *t,* is a regular feature of sound-development (Phonology, §§ 1 ; 13).

For the meaning of *Ashfield* see *Kirkby-in-Ashfield. Lownde* is from O.N. *lundr,* "wood" (see *Lound*). *Bonington* is the name of a separate parish (q.v.).

SYERSTON [saiəstn].

Type I.

1086 Sirestune, D.B.
1302 Syreston, F.A.
c. 1500 {Syerston Syreston} Inq. P.M. c. 1500.

Type II.

c. 1500 Syston, Inq. P.M. c. 1500.

O.E. *Sigerīces tūn,* "the farmstead of *Sigerīc.*" This latter is an O.E. man's name of frequent occurrence.

O.E. *īge* becomes M.E. *ī,* modern [ai] regularly. *č* is lost before *s* in an unstressed syllable (cp. *Ordsall*).

Type II shows loss of *r* before *s* (Phonology, § 13).

TEVERSAL.

Type I.

1086 Tevreshalt, D.B.
1227–77 Teversald, Non. Inq.
1284 Teversalt } F.A.
1316 Turessalt }

Type II.

1291 Tyv'salt, Tax. Eccles.
1346 Tyrvesalt } F.A.
1428 Tyvershalt }

The second element clearly stands for O.E. *holt,* "wood, copse." To this day the district can boast of an abundance of trees. In the unstressed syllable, *o* is unrounded (cp. *Egmanton*) and final *t*—having first become *d*—is dropped.

There can be no doubt that the first element is a pers. n. Prof. Skeat assumes that there existed an O.E. man's name *Tefer* (Pl. Ns. of Cambs.), whose first syllable, however, must have contained a long vowel, or a diphthong, $\bar{e}$ or $\bar{e}o$, for only from one of these can both the e and the i of Types I and II respectively be derived. The pl. n. *Tiverton* found in Cheshire and Devon may contain the same pers. n. It is possible that an assumed *Tēfer* represents the recorded O.E. *þeodfriþ* with the Norman initial t for *þ*. An O.H.G. *Tiufher*, in *Tiufherreshusun* (Förstemann), might be adduced here, although the initial consonants of the English and continental variants cannot easily be connected, unless we assume that t represents the Upper German variant of O.H.G. *d*, from *þ*.

THORESBY.

1086 Turesbi, D.B.
1316 Thuresby, F.A.

O.E. *þores bȳ(r)*, "the dwelling of *þor.*" The latter pers. n. which is of Norse origin occurs frequently both as *þor* and, less often, *þur*.

THORNEY.

Type I.

1086 Torneshaie, D.B.
c. 1500 Thorney, Inq. P.M. c. 1500.

Type II.

c. 1216 Thornehawe, Index.
1227–77 Thornhawe, Non. Inq.
1291 Thornhauwe, Tax. Eccles.
1302 Thornagh ⎰ F.A.
1316 Thorhawe ⎱

The two types differ in the form of the suffix: Type I contains O.E. *hege*, Type II O.E. *haga*, both meaning "hedge, enclosure." The name may, therefore, be translated by "thorn-hedge, or enclosure." The modern spelling shows influence of the more frequent suffix *-ey* from *ēge*, "island." The s of D.B. is due to false analogy: the clerk thought the first element was a pers. n. Similar cases of inorganic s are frequently met with in Anglo-Norman records (Zachrisson, pp. 118, 119).

THOROTON.

1086 $\left\{ \begin{array}{l} \text{Torvertune} \\ \text{Toruentun} \end{array} \right\}$ D.B.

1242 Thuruerton, Inq. P.M. I.
1284 Thorverton, F.A.
1363 Thoruerton, Index.
1637 Thoraton, Map in Camden.

"The *tūn* or homestead of *þurferð.*" This Scandinavian pers. n. occurs in various forms in English sources, such as *þurferð, þorfrð, Toruerd* etc. (Björkman, p. 155).

THORPE-IN-GLEBE.

1086 Torp, D.B.

$\left. \begin{array}{l} 1302 \\ 1535 \end{array} \right\}$ Thorp in Glebis $\left\{ \begin{array}{l} \text{F.A.} \\ \text{Valor Eccles.} \end{array} \right.$

The original simple name of D.B. times meaning "the village" had soon to be provided with a distinctive addition. The Latin *gleba*, English *glebe*, is used in its wider sense, meaning "a piece of cultivated land, field," as it still does in poetic language.

THORPE-BY-NEWARK.

$\left. \begin{array}{l} 1086 \\ 1153 \end{array} \right\}$ Torp $\left\{ \begin{array}{l} \text{D.B.} \\ \text{Index.} \end{array} \right.$

See preceding name.

THRUMPTON.

1086 Turmodestun, D.B.
1189 Turmodeston, Nottm. Ch.
c. 1240 Thurmunston, Woll. MSS.
1244 Thurmodeston, Cal. Rot. Chart.
$\left. \begin{array}{l} 1302 \quad \text{Thurmeton} \\ 1346 \quad \text{Thrumpton} \end{array} \right\}$ F.A.

"The *tūn* or farmstead of *þurmod.*" This pers. n. represents Scandinavian *þormóðr* on English territory. The spelling of the Woll. MSS. betrays influence of the pers. n. *þurmund.* In the course of development, the entire second syllable disappeared

with the exception of *m* between which and the following *t* a
labial glide arose (Phonology, § 16). Metathesis of *r* is frequently encountered in the pl. ns. of this county, and may take
place at any period (Phonology, § 15).

THURGARTON.

$$1086 \begin{cases} \text{Turgarstune} \\ \text{Turgastune} \\ \text{Torgartone} \end{cases} \text{D.B.}$$

c. 1170 Turgartona, Woll. MSS.
1189 Turgardton, P.R.
1278 Thurgarton, H.R.
1302 Thurgerton, F.A.

"*þurgār's tūn* or farmstead." The pers. n. is the English
form of the O.N. *þorgeirr*, with *ā* substituted for the cognate *ei*
(cp. *Tollerton*). It is remarkable that only the D.B. spellings
contain a genitival *s*. The P.R. form may have been influenced
by the feminine pers. n. *þurgerð*, the English variety of the O.N.
þorgerð r.

TILN or TYLN, TILNE.

$$1086 \begin{cases} \text{Tilne} \\ \text{Tille} \end{cases} \text{D.B.}$$

$$1189 \begin{cases} \text{Tilnea} \\ \text{Tilne} \end{cases} \text{P.R.}$$

1227–77 Tylne, Non. Inq.
1278 Tilne, H.R.
1293 Tylne, Index.
1535 Tilneye, Valor Eccles.

$$\begin{rcases} 1599 \\ 1637 \\ 1704 \end{rcases} \text{Tilney} \begin{cases} \text{Map.} \\ \text{Map in Camden.} \\ \text{Map.} \end{cases}$$

O.E. *æt Tilan ēġe*, " at *Tila's* island." This hamlet is situated
on the eastern bank of the river Idle. The complete loss of the
suffix is a remarkable feature (cp. *Blyth, Idle*).
Tila is an O.E. man's name.

TOLLERTON.

Type I.

1086 Troclauestune, D.B.
1166–7 Turlaueston, P.R.

Type II.

1294 Thorlaxton, Woll. MSS.
1302 Torlaxton, F.A.

Types I or *II* (continued).

1284 Torlaston ⎫ F.A.
1428 Toralston ⎭

c. 1500 ⎫ Torlaston ⎧ Inq. P.M. c. 1500.
1571 ⎭ ⎩ Index.

1578 Thorlaston, Index.

Type III.

1539 Torlaton, Bor. Rec.
1578 Torlarton, Index.
1637 Torloton, Map in Camden.

The second element is O.E. *tūn*, " farmstead." Two different pers. ns. seem to be contained in Types I and II respectively : *þorlāf* and *þurlac*. The former is an anglicised variety of O.N. *þorleifr*, whereas the latter stands for O.N. *þorleikr*. O.E. *ā* is frequently substituted for O.N. *ei*, see Björkman, p. 203 ; cp. *Thurgarton*. The pers. n. contained in Type I seems to be the original one, for which that of Type II was substituted. The forms arranged in the third section may have descended from either type, as *k* often disappears before *s*, and *v* is lost before a consonant.

Type III is the ancestor of the modern form. It is not probable that the *r* in the unstressed syllable represents the ending of the O.N. nominatives *þorleifr* or *þorleikr*. It is more likely that after the loss of *s*, *r* and *l* changed places, *rl* becoming *lr* (Phonology, § 15). The Index spelling of 1578 proves that the relative position of the two liquids was unsettled.

The D.B. spelling shows metathesis of *r* which is frequently found in that document (Stolze, § 29). I cannot explain the *c* satisfactorily.

The transition from initial þ to t took place under Norman-French influence. See Zachrisson, pp. 39 sqq., cp. *Teversal* and *Torworth*.

TORWORTH [toriþ].

> 1086 Turdenworde, D.B.
> 1278 Thorchewurh, H.R.
> 1316 Tordworthe, F.A.
> 1704 Tarworth, Map.

" The *weorþ* or habitation of *þoreð*, or *þureð*." This pers. n. goes back to O.N. *þorrǫðr* and is discussed at length by Dr Björkman (pp. 148 sqq.). *ch* and *h* in the H.R. seem to stand for *ð* and *þ* respectively. The spirant *þ* was occasionally mistaken for *h* by Norman scribes, but it would be difficult to explain why *ch* which usually denotes the sound of *k* before front vowels should represent *ð* in the H.R. spelling.

The loss of *ð* between *r* and *w* is natural. Initial *þ* was turned into *t* under Norman influence. See preceding name.

TOTON.

> *Type I.*
> 1086 Tolvestune, D.B.
>
> *Type II.*
> 1189 Turuerton, P.R.
>
> *Type III.*
> (*a*) 1086 Tovetune, D.B.
> 1284 } Toueton { F.A.
> 1314 } { Index.
> 1346 Touiton, F.A.
> 1480 Towton, Woll. MSS.
> (*b*) 1428 } Tauton { F.A.
> 1480 } { Woll. MSS.
> c. 1500 Tawton, Inq. P.M. c. 1500.

" The *tūn* or farmstead of *þorolf*." This pers. n. represents O.N. *þōrolfr*, and is found in an abbreviated form as *þolf*(*r*) both on Scandinavian and English territory. The latter variant constitutes the first element of Type I, from which Type III is derived. In Type II, the pers. n. *þurferð* seems to have been substituted for the original one.

The phonetic development is as follows: *l* became vocalised and the *v* soon disappeared after the diphthong *ou*. In the 15th century, M.E. *ou* and *au* seem to have become levelled under one sound, that of M.E. ɔ̄ (Phonology, § 9). This explains Types III *a* and III *b*, and also the modern spelling.

TRENT.

Trisantona, in Roman times.

Treanta, Bede, Hist. Eccles.

923   Treonta, A.S. Chron.

1278 Trent, Trenth, H.R.

A name of British origin, of which the meaning cannot be ascertained. See Bradley, Essays and Studies, I, p. 24; Miller, Pl. Ns. in the O.E. Bede, p. 52. Among early antiquarians the belief existed that this name was connected with Latin *triginta*, French *trente*, "thirty." This theory is set forth by Drayton in his "Polyolbion"; the passage deserves to be quoted. The poet represents the river as explaining its own name:

"In her peculiar praise, lo thus the River sings:
What should I care at all, from what my name I take,
That *Thirty* doth import, that thirty rivers make,
My greatness what it is, or thirty abbeys great,
That on my fruitful banks, times formerly did seat:
Or thirty kinds of fish, that in my streams do live,
To me this name of *Trent* did from that number give.
What reck I…" (26, 186 sqq.).

Milton alludes to the same interpretation in one of his earliest productions, when he sings:

"Of *Trent*, who like some earth-born giant spreads
His *thirty* arms along the indented meads."
                    (At a Vacation Exercise, 1627.)

TRESWELL or TIRESWELL.

Type I.

1086 Tireswelle, D.B.

1272–1307 Tyrswell, Index.

1278 ⎱
1302 ⎰ Tyreswell ⎰ H.R.
                  ⎱ F.A.

Type II.

1428 Tressewell, F.A.

$\left.{1535 \atop 1637}\right\}$ Truswell $\left\{{\text{Valor Eccles.} \atop \text{Map in Camden.}}\right.$

1704 Triswell, Map.

The meaning of the termination is clear: O.E. *wiell*, " fountain, spring." The first element is apparently a pers. n. It may represent an O.E. *Tīr, short for a full name composed with that element such as *Tīrweald*, *Tīrwulf*. There exists also an O.N. man's name *þyri* of which traces are found in English sources (Björkman, p. 164). Initial *t* for *þ* would be due to Norman-French influence (see *Tollerton*). Type II arose out of Type I through metathesis of *r* (Phonology, § 15). The discrepancy of vowels in Type II is not easily accounted for; the may have had some influence.

Both types survive in the alternative modern spellings.

TROWELL.

Type I.

$1086 \left\{{\text{Trowalle} \atop \text{Torwalle}}\right\}$ D.B.

c. 1200 Trowall, Woll. MSS.

Type II.

c. 1175 Trowella, Woll. MSS.

1227–77 Trouwell, Non. Inq.

1302 Trouell, F.A.

1637 Trowell, Map in Camden.

Type III.

1284 Trewell, F.A.

This is a difficult name to explain. I believe that the first element throughout the three types is O.E. *trēow*, " a tree, a forest; wood." The O.E. form of this word is most faithfully preserved in Type III, probably under the influence of the independent word, M.E. *tree*. In the other types, it has undergone certain changes caused by a shifting of accent in the triphthong *ēou*, which latter arose out of the vocalisation of *w* in

M.E. The development was as follows: $\bar{e}ow > \bar{e}ou > \acute{e}\bar{o}u > j\bar{o}u > \bar{o}u$, with loss of the glide j after r. It is impossible to say what the exact pronunciation of the combination ou, ow in the M.E. forms was. According to Camden's spelling, the contemporary pronunciation of the diphthong seems to have been the same as that of M.E. ou, au, ϱ, all three of which coincided in sound, as was shown under *Toton* (q.v.). If, however, the modern pronunciation [trauəl] is genuine, ow would stand for M.E. $\bar{u}$ ($<\bar{o}u$?).

The suffix is ambiguous, admitting of different interpretations. Type I, which seems to be the original, contains either O.E. *weall*, "a wall, rampart," or Scandinavian *vollr*, "a field, open country," in a more primitive form ($<$*valδuz*). For this ending the more usual *well* was substituted. If the former interpretation is accepted, the meaning of the pl. n. would be: "(at) the rampart made of wood, the palisade"; if the latter, one might translate by "the plain covered with trees."

It is unlikely that the second element was O.E. *weald*, "forest," although the early spellings do not preclude this, final *d* disappearing at a very early date after *l*. I cannot see what sense there would be in forming a tautology like *trēo weald*, "tree forest."

TUXFORD.

Type I.

1086 Tuxfarne, D.B.
1272–1307 Tuxforne, Inq. P.M. II.

Type II.

1258 Tuggesford, Inq. P.M. I.
1327–77 Tuxford, Non. Inq.
$\left.\begin{array}{l}1278 \\ 1316\end{array}\right\}$ Tukesford $\left\{\begin{array}{l}\text{H.R.} \\ \text{F.A.}\end{array}\right.$
1535 Tuxforde, Valor Eccles.

"The ford of *Tucca* or *Tuki*." The latter of these pers. ns. is of O.N. origin (Björkman, p. 142), the former is found in early Anglo-Saxon charters. Which of the two is implied in this case, it is impossible to say. The fact that the genitive ends in *es* and not in *an* might speak in favour of the Scandinavian name: yet examples of originally weak pers. ns. forming a strong genitive are by no means rare; see *Annesley*.

No importance is to be attached to the curious spelling, *gg* for *k*, of 1258.

I cannot explain the substitution of *n* for *d* in the suffix of Type I, unless it is due to a scribal error.

TYTHBY [tiðbi].

1086 Tiedebi, D.B.
c. 1190 Titheby, Woll. MSS.
1428 Tythby, F.A.
1535 Teythby, Valor Eccles.

The suffix is the well-known Scandinavian *bȳ(r)*, "dwelling." The first element may be a pers. n., but I am unable to suggest what its exact form and derivation were. The spelling *ie* in D.B. is remarkable and seems to imply that the vowel intended to be represented was M.E. *ē*, whatever its source; see Stolze, p. 9.

UPTON.

1086 $\left\{ \begin{array}{l} \text{Upetun} \\ \text{Opeton} \end{array} \right\}$ D.B.

This village is said to be situated "on a gentle acclivity" (White, Directory, 1853, p. 520). See following name.

UPTON (in Headon Parish).

1086 Upetone, D.B.
1278 Upton, H.R.
c. 1500 Upthorp, Inq. P.M. c. 1500.

"The *tūn* or farmstead on high ground." This second Upton occupies the highest part of the surrounding country. The prefix *ūp* is used in O.E. to denote a high situation, as in *ūp-hūs*, "upper chamber," *ūp flōr*, "upper floor," etc. It is also encountered in numerous pl. ns. other than *Upton*, as in *Upminster*, *Upwood*. The same element has a similar meaning in the Low German pl. ns. *Uphausen*, *Upstede* etc., Jellinghaus, p. 325.

The substitution of *-thorp* for *-ton* does not call for an explanation.

WALESBY.

$\left.\begin{array}{c} 1086 \\ 1204 \end{array}\right\}$ Walesbi $\left\{\begin{array}{l} \text{D.B.} \\ \text{Index.} \end{array}\right.$

1316 Walesby, F.A.

"The *bȳ(r)* or dwelling of *Wealh*, or the Briton, or the slave, serf." The original meaning of O.E. *wealh* is "foreigner, Briton, Welshman"; after the subjection of the Britons, the word assumed the sense of "slave, serf." It also occurs as a pers. n. In which of these three senses the word is used in the above pl. n. it is impossible to say.

WALKERINGHAM.

1086 Wacheringeham, D.B.

c. 1216 Walcringham

$1272-1307 \left\{\begin{array}{l} \text{Waucringham} \\ \text{Walcringham} \end{array}\right\}$ Index.

$1278 \left\{\begin{array}{l} \text{Waleringham} \\ \text{Waveringham} \\ \text{Walcringham} \end{array}\right\}$ H.R.

1291 Waltryngham, Tax. Eccles.

1316 Walcringham, F.A.

1637 Walkingham, Map in Camden.

"The home of the descendants of *Walchere*," O.E. *Wealh-heringa hām*. The O.E. pers. n. *Wealhere* frequently appears as *Walchere* (*ch*=[k]). This same patronymic is probably contained in the Yorks. pl. n. *Walkingham* (Pl. Ns. of the W. Rid.).

The early forms exhibit a considerable variety of spellings which are, however, of but small importance. *au* for *al* shows Norman vocalisation of *l* (cp. *Mansfield*; Zachrisson, pp. 146 sqq.). *t* for *c* = *k* is a scribal mistake frequently met with in mediæval documents.

WALLINGWELLS.

Type I.

1278 Wellandwell, H.R.
1289 Wallendewelles, Inq. P.M. II.
1291 Wallandwells, Tax. Eccles.
1300 Wallandewelles, Index.

Type II.

1516 Wallingwells, Bodl. Ch. and R.
1637 Woldingwells, Map in Camden.

O.E. *weallende welles*, "the boiling, i.e. bubbling and flowing springs." This is the site of a former nunnery which is described by the pious founder as " [unus locus] in meo parcho de Carletuna juxta fontes et rivum fontium "—by the wells and the stream of the wells (Dugdale, Monast. Angl., new ed. IV, 295, temp. reg. Steph.).

The first part of this name is a regular O.E. present participle. The change from *en* to *ing* took place at a comparatively late date; cp. Types I and II. I am not prepared to say whether this transition is due to a special sound-law (Phonology, § 13) or to analogy of the verbal nouns which always ended in *-ing* as assumed by Dr Sweet, N. Engl. Gramm. § 1239.

We find a corresponding O.H.G. name in *Wallendenbrunno*, "at the flowing well or spring" (Förstemann, II), modern Walbernbrunnen, Odenwald.

WARSOP.

$\left. \begin{array}{l} \text{Wareshope} \\ \text{Waresope} \\ \text{Warsope} \end{array} \right\}$ D.B.

1272–1307 Warsop *alias* Warshope, Inq. P.M. I.
1302 Warsop, F.A.

O.E. *Wares hop*, "the valley of *Ware.*" The latter male name is recorded but once in the Onomasticon. It may have been more frequent as a short form of a name composed with *Wær-*, such as *Wærbeorht, Wærmund* etc. A similar name is contained in the Low German pl. ns. *Warenrode, Warantharpa*, modern Wahrendorf near Münster (Förstemann, I).

WATNALL [wotnə] (Watnall Chaworth and Watnall Cantelupe).

Type I.

1086 Watenot, D.B.

Type II.

c. 1200 Wattenhou, Woll. MSS.
1216–1307 Watenhow, Testa de N.
1278 Watenhou, H.R.
$\left.\begin{array}{l}1302\\1316\\1346\end{array}\right\}$ Watenowe, F.A.
1506 Watnowe, Bor. Rec.
1535 Whatnaw, Valor Eccles.

Type III.

c. 1700 Watnall, Map in Camden (ed. 1722).

"The hill of *Wata*," O.E. *Watan hōh.* The final *t* in the D.B. spelling is an error for *c*, which stands for *h* (cp. *Wysall*; see Zachrisson, Latin Influence, p. 22). Type I contains the suffix in the nominative, whereas Type II goes back to the dative *æt Watan hōȝe.* The diphthong *ow* from *ōȝ* must have become the same in pronunciation as M.E. *au* with which it was confused by the compiler of Valor Eccles. This *au* or *aw* in many cases stood for older *-al*, from *-healh*, *-hale*, and it was for that reason that the suffix came to be written *-all*, as in Type III and the modern spelling, although the *l* was never pronounced, except by those who rely on the spelling only.

The element *hōh* is treated at length by Prof. Wyld in Pl. Ns. of Lancs. p. 351. Its meaning in O.E. is "heel, hill, promontory." Watnall is situated on an eminence from which it apparently derives its name.

Cantelupe and Chaworth are the names of former owners belonging to the Norman nobility.

WELBECK (Abbey).

c. 1189 Wellebec, Nottm. Ch.
1278 Wellebeck, H.R.
$1284\left\{\begin{array}{l}\text{Welbek}\\\text{Welbecke}\end{array}\right\}$ F.A.

The suffix is the Scandinavian *bekk(r)*, "a brook." The first element apparently stands for O.E. *wiell, well,* "stream, spring."

The meaning of the compound seems to be "the brook flowing from the spring."

WELHAM.

$$1086 \left\{ \begin{array}{l} \text{Wellun} \\ \text{Wellon} \end{array} \right\} \text{D.B.}$$

1276 Wellum, Index.
1316 Wellom, F.A.
1457 Wellum, Index.

O.E. *æt wellum*, "at the wells, or waters." See preceding name, and cp. *Kelham*. The name is explained by White (Directory, 1853, p. 693) as follows: "Its name is derived from *St John's well*, which was long famed for its medicinal virtues in scorbutic and rheumatic complaints; it is now a commodious *bath*, though it has lost much of its former celebrity."

WELLOW.

$$1278 \left\{ \begin{array}{l} \text{Welhagh} \\ \text{Welhah} \end{array} \right\} \text{H.R.}$$

1302 Welhawe, F.A.
1535 Wellaw, Map in Camden.
1704 Welley, Map.

The suffix stands for O.E. *haga*, "a hedge, fence; hence a piece of ground enclosed or fenced in, an enclosure." A word of similar meaning is O.E. *hege* which may be contained in the spelling of 1704, although it is more likely that -*ey* represents the more usual termination, O.E. *eğe*, introduced by the engraver.

The first element looks like O.E. *wiell*, *well*, "a spring, stream." However, I cannot say what is the exact meaning of the compound.

For the spelling *ow* instead of M.E. *aw* see Phonology, § 9.

WESTON.

1086 Westone, D.B.
1268 Wiston, Index.
1302 Weston, F.A.

The meaning is obvious. If the spelling *Wiston* is genuine, the change from *e* to *i* must be explained as caused by the following *s*.

WHATTON.

1086 Watone, D.B.

$\left.\begin{array}{l}\text{1284 Watton}\\\text{1302 Whatton}\end{array}\right\}$ F.A.

O.E. *hwǣte tūn*, " the wheat-enclosure, or the farm near which wheat grows." The vowel of the first syllable underwent early shortening before *tt*: *ǣ* > *æ* > M.E. *a*. See the following name.

It has been suggested that the name is derived from *water* on account of the watery situation of the village, or from *wathtun*, *wath* being a Scandinavian word for "ford," because there still exists in the locality a ford across the river Smite. But both these interpretations fail to account for the initial *wh* which cannot be merely a late fanciful spelling.

WHEATLEY.

$1086\left\{\begin{array}{l}\text{Wateleie}\\\text{Wateleia}\\\text{Watelaie}\end{array}\right\}$ D.B.

$1278\left\{\begin{array}{l}\text{Wetele}\\\text{Wetelay}\end{array}\right\}$ H.R.

$\left.\begin{array}{l}\text{1302 Westley}\\\text{1316 Whetleye}\end{array}\right\}$ F.A.

$1327\text{--}77\left\{\begin{array}{l}\text{Weteley}\\\text{Whiteley}\end{array}\right\}$ Non. Inq.

O.E. *in hwǣte lēaȝe*, "in the wheat field." *a* in D.B. stands for O.E. *ǣ*, M.E. *ē*. *Westley* and *Whiteley* are spellings obviously caused by false etymology.

WIDMERPOOL [locally: windəpul; otherwise: widməpūl].

$1086\left\{\begin{array}{l}\text{Wimarspol}\\\text{Wimarspold}\end{array}\right\}$ D.B.

1189 Widmespol, P.R.

$\left.\begin{array}{l}\text{1284 Witmerpol}\\\text{1428 Wodemerpole}\end{array}\right\}$ F.A.

1571 Widmerpole, Index.

O.E. *Wīdmǣres pōl*, "the pool or pond of *Wīdmǣr*." The loss of the genitival *s* is a remarkable feature of the development of this name. Popular etymology is responsible for various interpretations embodied in the early spellings. The second syllable of the pers. n. was taken for *mere*, "lake, pond," and the first was connected with the adjective *white* (F. A. 1284), and the noun *wood*. Two different explanations are offered by Thoroton, who thinks that the name means either " Wimears Poll or Spear, or Wide mere Poole" (I, 77).

The metathesis of *dm* 'in the local pronunciation is unusual : *dm* became *md*, the *m* being afterwards turned into *n* through assimilation.

The final *ld* of the second D.B. form is puzzling. It is not impossible that it stands for *dl*, the suffix being *pudel*, *puddle* instead of *pōl*. Metathesis of *d* is not infrequent in D.B. (Stolze, § 30).

WIGSLEY or WIGGESLEY.

> 1086 Wigesleie, D.B.
> 1260 Wiggesley, Index.
> 1302 Wyggesleye, F.A.

The suffix is O.E. *lēah*, "a field." The pers. n. involved may be either O.E. *Wicga* or Scandinavian *Vīgi*. The former is found in O.E. sources ; the latter occurs in Scandinavian records (Rygh, G. Personnavne, s.v.).

WIGTHORPE.

No early forms. The meaning of the second element is clear. The first element may be either of the pers. ns. mentioned under the preceding name.

WILFORD.

> 1086 Wilesforde, D.B.
> 1184–1204 Wileford, Woll. MSS.
> 1302 Wilford, F.A.

Probably O.E. *Willan ford*, "the ford of *Willa*, leading to *Willa's* habitation." The *s* of D.B. is spurious (cp. Zachrisson, pp. 118, 119).

In White's Directory (1853, p. 405) the following note on the etymology of this name is found: "The church...is dedicated

to St Wilfrid, and the name of the village is evidently a con-
traction of Wilfrid's ford, as there is both a ford and a ferry
close by." This statement confuses cause and effect; for if
there exists any connection between the patron saint and the
name of the village, the former must have been chosen in order
to suit the latter.

WILLOUGHBY.

$\left.\begin{array}{l}1086 \\ 1156\end{array}\right\}$ Wilgebi $\left\{\begin{array}{l}\text{D.B.} \\ \text{Index.}\end{array}\right.$

1204 Wilghebi, Index.

1291 Wilweby, Tax. Eccles.

$\left.\begin{array}{l}1302 \quad \text{Willeugby} \\ 1428 \quad \text{Welughby}\end{array}\right\}$ F.A.

See following name.

WILLOUGHBY-ON-THE-WOLDS.

1086 $\left\{\begin{array}{l}\text{Wilgebi} \\ \text{Willebi}\end{array}\right\}$ D.B.

c. 1180 Wilghebi, Woll. MSS.

1252 Wiliugeby super Wolde, Inq. P.M. 1.

1363 Wilughbi super Waldas, Index.

"The $b\bar{y}(r)$ or dwelling among the willows." The suffix is
of Scandinavian origin. The first element corresponds to modern
English *willow*, whose M.E. ancestor is *wilow*, *wilwe*. In O.E.,
this word is found as *weliȝ*, *wyliȝ*, which normally would result
in M.E. *wily*. We are, therefore, forced to assume that there
existed in O.E. a variant containing a back vowel in the second
syllable, **welug*, *wylug*, ending in the back-open instead of the
front-open consonant. The difference in the unstressed vowels
is to be ascribed either to gradation (Sievers, §§ 127, 128), or to the
circumstance that the two nouns belonged to the *o* or *jo* classes
of strong masculines respectively. The mutated vowel of the
one might have been transferred to the other.

The first named Willoughby is situated among low-lying
meadows on a brook near which willows flourish abundantly.

WIMPTON.

1086 Wimunton, D.B.
1168–9 Wimunttun, P.R.
1316 Wympton, F.A.
c. 1500 $\left\{ \begin{array}{l} \text{Wynton} \\ \text{Wympton} \end{array} \right\}$ Inq. P.M. c. 1500.

The P.R. spelling is the most conclusive. This is not the "women's *tūn*," but "the *tūn* or farmstead of *Winemund*." This O.E. man's name is recorded once in the Onomasticon, as borne by an "amicus" of Eadbeald, king of Kent. Cp. *Winthorpe*.

n became assimilated to the following *m* at a very early period. A similar fate happened to the *m* at a later date when it came into contact with the dental *t*, which changed it to *n*; see *Wynton* (Inq. P.M. c. 1500). This form seems to have perished. Where the *m* was retained, a glide developed between it and the following *t* (Phonology, § 16). The absence of the genitival *s* is noteworthy.

WINKBURN or WINKBOURNE.

1086 Wicheburn, D.B.
1153 Winkeburn, Index.
1189 $\left\{ \begin{array}{l} \text{Winkerburn} \\ \text{Wincheburn} \end{array} \right\}$ P.R.
1278 $\left\{ \begin{array}{l} \text{Winckeburne} \\ \text{Wingeburne} \end{array} \right\}$ H.R.
1346 Wynkeborn, F.A.
1637 Winkborn, Map in Camden.

The etymology of this name is very doubtful. The inquiry is rendered more complicated by the fact that the little river flowing past the village is called *Wink* on modern maps, and *Winkle* by older writers, e.g. in White's Directory of 1853 (p. 521).

The suffix is O.E. *burne*, "spring, brook." The whole name originally applied to the water-course, the habitation or village on its banks being called "at Winkburn." The first element may have been an O.E. adjective **wincol*, derived from the root contained in *wincian*, "to wink, blink," and O.H.G. *winchan*, whose original sense was "to bend." Similar adjectival formations are numerous in O.E.; an alternative ending is *-er*, *-or* which is occasionally interchangeable with *-ol*, as in *wacol*, *wacor*, "vigilant." The meaning of O.E. **wincol* would be "winding, pliant," which might very well be applied to many brooks. The variant **wincor* would account for the first P.R. spelling, unless the first *r* stand for *l*, as these two liquids are frequently interchanged in records compiled by Norman scribes.

The river-name *Winkle* can be explained as a back-formation from *Wincel burn*, "the winding brook." The modern name *Wink* probably came into existence by the same process but at a later date, when *l* had disappeared in the pl. n. By a similar method the river-name *Maun* was deduced from the town *Mansfield* (q.v.).

WINTHORPE.

1086 Wimuntorp, D.B.

1165–1205 {Wimethorp / Wimetorp / Winetorp} Index.

1291 Wymthorp, Tax. Eccles.

1316 Winthorp, F.A.

"The hamlet of *Winemund*." The same pers. n. forms the first element of *Wimpton* (q.v.).

WISETON or WYESTON [wistn].

1086 Wisetone, D.B.

1278 Wiston, H.R.

1304 Wystone, Index.

1316 Wyston, F.A.

It is clear that the ending represents O.E. *tūn*, "homestead," and that the first element is a pers. n. O.E. *Wīsa* is not quoted in the Onomasticon as having been borne as a name by any known person. We may, however, safely assume that it existed. The meaning of *Wīsa* is "the wise one" or "the leader." O.E. *Wīsan tūn* seems to have developed in two different directions.

The *n* was lost in both cases according to a general rule. The *e* of the second syllable was either retained or dropped. In the former case the *ī* stood in an open syllable and remained long as in the modern spellings; in the second case, *ī* before *st* was shortened, giving rise to the modern pronunciation.

WITHAM (river).

The suffix may be an O.E. *amma*, a somewhat doubtful name for a river, probably of Celtic origin, see Middendorf, s.v. The first element may represent O.E. *wīþig*, "willow." This conjecture receives further support from the fact that the banks of the above river must at one time have been famous for the abundance of osiers, as the village of Barnby situated on it has received the distinctive addition of "in-the-willows." Thus the meaning is "the willowy stream." Names like *Weidenbach* of the same meaning are of frequent occurrence throughout Germany.

WIVERTON [wᴀ̄tn].

1086 Wivretuɳe, D.B.
c. 1180 Wiverton, Woll. MSS.
1284 Veverton ⎱ F.A.
1302 Wyverton ⎰
1637 Waerham, Map in Camden.
1704 Waerton, Map.
Wiverton (commonly called) ⎱ Thoroton, I, 195.
Werton ⎰

The first element of this name must be a pers. n., either O.E. *Wigferþ*, *Wigfriþ*, or *Wīdfara*, the last of which is claimed to be of Scandinavian origin by Dr Björkman. It occurs in a shortened form as *Wiuare*, *Wifare* in D.B., which exactly corresponds to the early spellings above. The suffix is, of course, O.E. *tūn*, "homestead"; Camden blunders by substituting *-ham*. The absence of any sign of a genitival ending need not disturb us. After the loss of *v*, *ir* developed in the regular way (Phonology, § 8).

WOLDS (a range of hills).

1252 Wolde, Inq. P.M. I.

1363 Waldas, Index.

1613 {(the large, and goodly full-flocked) oulds } Drayton's Polyolbion.

From O.E. (Mercian) *wáld*, "forest." *a*, which had become lengthened before *ld*, was changed into *ǭ* in M.E. Drayton's spelling represents a dialect pronunciation, showing loss of initial *w* before a rounded back vowel (cp. *ooze* < M.E. *wose*, Horn, § 173).

The above word is identical with Southern English *Wealds* which is descended from the W. Saxon and Kentish variant *weald*. The development of senses can be easily traced. When the plains had been practically cleared of woods and forests, the word *weald, wald* was gradually restricted to the hills still covered with trees.

WOLLATON [wulətn].

1086 { Ollavestone Olavestune Waletone } D.B.

1216–1307 Wullaveton, Testa de N.

1284 Weleston 1302 Woleton } F.A.
1316 Wolaton

1327–77 Wolaston, Non. Inq.

1428 Willaton, F.A.

The spelling of Testa de N. is the most helpful; if it contained the *s* found in some others it would be perfect. The O.E. prototype was *Wulflāfes tūn*, "the homestead of *Wulflāf*." This pers. n. occurs frequently in O.E. documents. The loss of the genitival *s* is noteworthy.

The D.B. scribes found it impossible to render the sound of *wu* and often blundered over it (see following name). The spellings containing *a, e* or *i* in the first syllable are clerical mistakes and require no explanation.

WOODBOROUGH.

$$1086 \left\{ \begin{array}{l} \text{Udeburg} \\ \text{Udesburg} \end{array} \right\} \text{D.B.}$$

1278 Wodeburg, H.R.
1302 Wodeburgh, F.A.
1637 Woodbro, Map in Camden.

" The fortified place in the wood," from O.E. *wudu*, " wood," and *buruh*, " fortified place." The development is regular. The *s* in the second D.B. spelling is spurious; the scribe probably took the first element for a pers. n. Camden records the contemporary pronunciation.

WOODCOTES.

1302 Wodicotes, F.A.

" The cottages in the wood," O.E. *þā wudiȝe cotas*. The second syllable of the adjective *wudiȝ*, " woody," has been lost altogether.

WOOLLEN.

I suggest that the suffix is the same as in *Meden*, and *Witham* (qq.v.). If so, the first element might be O.E. *wulf*, " wolf," and the meaning of the compound " wolf stream." In O.H.G. a corresponding *Wolfaha* is found, modern *Wolfach* (Förstemann, II).

WORKSOP [wasəp] (Worsop, Wossap, Wursup, Hope).

Type I.

$$1086 \left\{ \begin{array}{l} \text{Werchesope} \\ \text{Werchessope} \end{array} \right\} \text{D.B.}$$

1189 Werchessope, P.R.

Type II.

1189 Worcheshope, P.R.
1410 Worsope, Index.

Type III.

1302 Wirksop, F.A.
1327–77 Wyrksop, Non. Inq.

1345 Wirkesop, Index.
1346 Wirsop, F.A.

Type IV.

1535-43 Werkensop, Leland, I, 99.
1637 Workensop, Camden, p. 550.

The termination in Types I, II, III apparently is O.E. *hop*, "valley." The first element is a pers. n., of which there is no record in the Onomasticon, it is true, but whose existence is proved by its being contained in a number of pl. ns. enumerated by Prof. Moorman, s.v. *Worksborough*. That name seems to have occurred in various forms, namely as *Wyrc* (Type III), *Weorc* (Type I), and *Worc* (Type II). The last form evidently developed out of the second on or near Northumbrian territory, where the change from *weo* to *wo* took place (Bülbring, § 265).

The modern spelling goes back to Type II, so do the first and the second pronunciations recorded by Hope. The pronunciation in square brackets is descended from either Type I or III, *ir*, *er* first becoming [ā] as in the last of Hope's forms, and then being shortened.

Type IV is puzzling. The two spellings are too late and can hardly be taken seriously. It is, moreover, very probable that Camden partly copied from Leland. One might feel tempted to say that the *n* represents the genitival ending of a weak variant of the pers. n. *Weorca*; but the presence of an additional *s* renders that impossible.

WYSALL [waisə].

Type I.

1086 Wisoc, D.B.

Type II.

1302 Wishow, F.A.
1327-77 Wysowe, Non. Inq.
1428 Wyshow, F.A.
1476 { Vyssow } Index.
        { Wysow }
1535 Wyshawe, Valor Eccles.
1637 Wysshaw, Map in Camden.

The suffix is the same as in *Watnall* (q.v.); D.B. writes final *c* instead of *h* in *hōh*, "hill." The nature of the prefix is extremely doubtful. It might stand for the pers. n. *Wīsa*, found in *Wiseton*. If so, the meaning of O.E. *Wīsan hōh* would be " *Wīsa's* hill."

The development of the suffix and the recent substitution of *-all* for the ending are explained under *Watnall*.

If the explanation offered is correct, one would expect the *s* to be voiced. I cannot satisfactorily account for the quality of that consonant.

In the spellings of 1535 and 1637, *shaw*, from O.E. *sceaga*, " small wood, copse, thicket," is substituted for the original suffix. These forms are too recent to be classified as a genuine type.

PHONOLOGY OF THE NOTTINGHAMSHIRE PLACE-NAMES

N.B.   Only those sound-changes that are of special interest have been classified here.

I. VOWEL CHANGES.

§ 1.   Shortening of original long vowels in stressed syllables before certain consonant combinations.

(a)   Late O.E. shortenings :

ā > a: Aslockton < Āslac ; Bradmore < brād- ; Stanton, Stanford < stān.

ǣ > a : Clareborough < clǣfre- ; Martin < mǣr- ; Whatton < hwǣt- ; Hatfield < hǣþ-.

ēa > a : Radford, Ratcliffe < rēad-.

ē > e (W. Sax. ǣ, Mercian ē): Fledborough < Flēda ; Strelley < strēt- ; Sturton, older Stretton < strēt-.

ēo > e : Darlton < *derl- < Dēorl-.

ī > i : Limpool, Linby < līnd-.

ō > o : Broxtow < Brōcwulf- ; Hokerton < Hōkr : Ossington < Ōskin-.

ū > u : Dunham < dūn- ; Rufford < rūh- ; Southwell [saðl], Sutton < sūð- ; Plumtree < plūm.

(b)   Early M.E. shortenings:

ēa > e : Edwalton, Edwinstow < Ēad- ; Retford < rēad-.

ē > i : Gringley < grēn ; but see § 6.

§ 2.   The regular development of O.E. ā is M.E. ǭ, modern [ou]: Gotham < gāt- ; Goverton < Gārfrið- ; Grove < grāfe.

§ 3.   M.E. ih seems to have developed into late M.E. ī which did not participate in the diphthongisation of early M.E. ī from O.E. ī, ȳ, iʒ: Kneesall, Kneeton < kniht- (?).   In the Dialect Grammar (§ 77) it is stated that this change has taken place in Northumberland, Durham, Cumberland, Westmoreland, Yorkshire, Lancashire, Cheshire, Flintshire, and parts of Staffordshire, Derbyshire, and Lincolnshire.   As Nottinghamshire is situated between Yorkshire, Derbyshire, and Lincolnshire, it is not unreasonable to assume that the same sound-change has taken place within its territory. The Nottinghamshire dialect is very badly represented in the Dial. Dict. ; I am informed that old people still pronounce night as [nīt], but that otherwise the standard pronunciation has supplanted the genuinely local forms. See Horn, § 69 anm.

§ 4. M.E. ū is derived from various sources and has developed on various different lines. (1) O.E. ū has undergone diphthongisation in: Lound < *lūnd*, but was shortened in: Southwell [saȫl] < *sūþw-*. (2) After *u*, *l* and *h* became vocalised; the resulting combination *uu* was then treated as older *ū*, i.e. diphthongised to [au]. The first change has taken place in: Boughton < *buhtūn*, and Bulcote < *bul-*, but *ū* was not diphthongised through influence of the initial labial consonant; cp. the pronunciation of *wound*.

M.E. *ū* from *ul* has become shortened at a later period in: Sowlkholme [saᴋm] < *sulh-*.

In Oldcoates < *ūle-*, M.E. *ū* seems to have developed into modern [ei], the pronunciation of that name being given as *Alecotes* in Hope's Glossary. This may represent the genuine dialect pronunciation, for which, however, I cannot find any confirmation.

§ 5. Early M.E. **æ**, **a** appears as both *e* and *a* before original *ks*: Laxton, older Lessington < *Leax-*; cp. Dial. Grammar, § 25.

æ seems to have been lengthened before *s* in: Basford [beisfəd] < *Bæssa-*; this M.E. change corresponds to a similar lengthening of *æ* before open voiceless consonants in modern southern English, in words like *mass*, *grass*, *path*.

§ 6. M.E. **e** has become *i* before *ng* [ŋ]: Bingham < *Benninga-*; Finningley < *fenninga-*. Cp. Dial. Gramm. § 55, Horn, § 38.

§ 7. M.E. *er* and *ar* developed into *ær*; after this change had taken place, the *r* was lost before the dental and blade consonants *s*, *þ*, *l*, *n*, the *æ* being lengthened at the same time: *ers*, *ars* > *ærs* > *ǣs*. This *ǣ* is represented by modern [ei], or [æ] if shortened: Bassetlaw, Caythorp, Dalington, Danethorpe, Perlethorpe, older Palethorpe, Staythorpe: see discussions and forms under each name above.

§ 8. M.E. **er**, **ir**, and **ur** have all become [ʌ] in the modern pronunciation; in some cases, however, *er*, *ær* are represented by *ar* and [ei]; see § 7. The early spellings of some of the following place-names prove that this transition took place at a somewhat earlier date than is usually assigned to it; see Barnston, Barton, Carburton, Darlton, Girton, Harby, Serlby, Sherwood, Sturton, Wiverton, Worksop.

§ 9. M.E. **au** and **ou**. The combination *al* developed into *aul*; the *l* is retained in the modern spelling of: Balderton, Calverton [kɔ̄vətn, kāvətn], and Halloughton [houtn, hɔ̄tn]; it has been lost in: Caunton < Caln-, and Awsworth < Aldes-. M.E. *ǭ* < *au* was shortened in: Ollerton and Ompton < *alr-*, *alm-*.

ol has become [oul] in: Rolleston; it is represented by [ɔ] in Hawton < *holt-*. This latter fact seems to prove that, in the dialect, M.E. *au* and *ou* have fallen together. The spelling *Hauton* for older *Holton* is found as early as 1270.

Scand. *ou* appears as *ō*, *o* in: Haughton, Hockerton; cf. Björkman, Scand. Loan Words.

§ 10. Shifting of accent is found in: Keyworth and Trowell (qq.v.).

§ 11. Norman influence accounts for the change from *an* to *aun* in: Maun, Saundby, Staunton.

II. CONSONANT CHANGES

§ 12. **Loss of Consonants.** Of three consonants, the middle one is lost, this change being due to assimilation in most cases.

d : Bilsthorpe < -*lds*- ; Bonbusk < -*ndb*- ; Chilwell < -*ldw*- ; Felley < -*ldl*- ; Shelford < -*ldf*-.

f : Wollaton < -*lfl*-.

k : Kirton < -*rkt*- ; Radcliffe [rætlif] < -*tkl*- ; Syerston < -*rks*- ; Worksop [wasəp] < *rks*-.

l : Gamston < -*mlst*-.

n : Milton < -*lnt*-.

t : Beeston [bīsn] < -*stn*-.

þ, ð : Normanton < -*rðm*- ; Norney < -*rðrn*- ; Norton < -*rþt*- ; Norwell, Torworth < -*rðw*- ; Wiverton < -*rþt*-.

v : Elston < -*lvst*- ; Scarle < -*rvl*- ; Selston < -*lvst*-.

§ 13. **Assimilation.**

ld > *ll* : Bole, Bolham, Rolleston.

dk > *kk* > *k* : Eacring < $\bar{E}ad\langle wæ\rangle cer$-.

fn > *mn* : Rampton < **Hrafntūn*.

hf > *ff* : Rufford < *rūhford*.

sk > *ss* : Ossington < $\bar{O}skintūn$; Bothamsall < -*skeld*.

ks > *ss* : Laxton, older Lessington.

kþ > *þ(þ)* : Huthwaite < -*kþw*-.

kt > *tt* : Boughton (Button, Type III) < *buktūn*.

tw > *kw* : Bestwood, older Beskwood.

ms > *ns* : Mansfield < *Mams*-.

nmnþ > *mþ* > *nþ* : Winthorpe < *Winemundþorp*.

nŋ > *ŋ(ŋ)* : Bingham < *Benning*-.

nm > *m(m)* : Wimpton < *Winmun*- ; Kimberley < *Cynmǣr*-.

ndþ > *nþ* > *mþ* : Limpool < *lindþōl*.

ŋt > *nt* : Dalington [dælintn].

stw > *skw* : Bestwood, older Beskwood.

ts > *ss* : Cossal < *Cots*-.

tl > *dl* > *ll* : Strelley < *strētleȝe* ; Teversal < -*holt*.

rs > *s(s)* : Syerston (Type II).

rð > *rr* : Scarrington < -*rð*-.

þt > *tt* : Sutton.

b—n > *b—m* ("Fernassimilation," see Horn, § 228 anm.): Bramcote < *bran(d)cot* ; Brinsley, see Testa de N. spelling.

n in the second syllable of trisyllabic words has become ŋ in a number of place-names. This second syllable had very weak stress, and I assume that this vowel, of whatever origin, became *ĭ* at a very early period. This high front vowel exercised an assimilatory influence upon the *n*, changing the dental nasal into the front nasal (retracted). A similar change is observed in the M.E. present participle *bīndinde* becoming *bīndinge*. This transition is usually ascribed to the influence of the corresponding verbal nouns in *-inge* (Sweet, N. Engl. Gramm. § 1239), but it may have been assisted and accelerated by the operation of the sound-law formulated above. Other writers on place-names assume that the change from *-an-*, *-en-*, *-in-* to *-iŋ-* is due to analogy with names containing an original patronymic particle *-ing-* (Wyld, Place-Names of Lancashire, p. 36, see Alexander, Essays and Studies, II, pp. 158 sqq.). There are two circumstances which speak against such an explanation : (1) The large number of *-ing-* forms derived from *-in-* (Alexander, l.c. p. 181). (2) The existence of modern vulgar pronunciations exhibiting a similar transition in independent words, as *skelington* (skeleton, older *skelenton, cf. celandine < Latin *celidonia*, etc., Horn, § 225), and *sartingly* (certainly). Examples of Nottinghamshire place-names are : Edingley < *Ēadwin-* ; Attenborough (see older forms); Farndon (see older forms); Kilvington (?); Kirklington (?); Laxton, older Lessington < *Leaxan-*; Newington < *nīwantūne*; Ossington < *Ōskin-*; Wallingwells < *wallend(e)-*.

§ 14. **Dissimilation.** Loss of *r* through influence of another *r* contained in the same word is found in : Goverton, and perhaps in : Ordsall (qq.v.). The loss of *n* in Misterton < *mynstertūn* may be due to the dissimilatory effect of the initial *m*. (Cp. Zachrisson, pp. 136 sqq.; Horn, § 237.)

§ 15. **Metathesis.**

r frequently changes its place in the accented syllable : Girton < *grēot-* ; Scrooby < *scurva-* ; Sturton < *strēt-* ; Thrumpton < *þurmōd-* ; Treswell < *tires-*.

n(m) changes its position in : Widmerpool [windəpul] < *Wīdmǣr-*.

§ 16. **Development of a glide.**

mt > *mpt* : Ompton, Plumtree, older Plumptre, Rampton, Thrumpton Wimpton.

mst > *mpst* : Rempston.

mr > *mbr* : Kimberley.

pl > *pwl* : Cropwell.

§ 17. **Voicing and unvoicing**, due to partial assimilation.

dk > *tk* : Ratcliffe < *rēad-*.

df > *tf* : Retford < *rēad-*.

hb > *p* : Epperston < **Eohberht-* (?).

tb > *db* : Budby < *But-*.

§ 18. **Loss of h and w** at the beginning of an unstressed syllable.

h is lost : Cossal < *-hale*; Cropwell < *-hill*; Nottingham [notiŋm], etc.

w is lost : Colwick [kolik]; Bulwell [buləl]; Eakring < *Ēadwæcer-*; Edwinstow, or Edenstow; Harworth [hærəþ]; Kinoulton < *Cynweald-*; Norwell [norəl]; Southwell [saᵭl].

§ 19. Initial *h* is a very unstable element in the Nottinghamshire dialect. It is usually dropped in pronunciation, but may be prefixed to any stressed word beginning with a vowel : Appesthorpe, Hickling, Hockerton, Hoveringham.

§ 20. **kt > ht.**
Whenever in the older Germanic languages *k* and *t* met in combination, the former was opened and changed into the back or front spirant. This same sound-law seems to have been in operation throughout the O.E.[1] and M.E. periods, unless crossed by analogy. Prof. Wyld was the first to draw attention to this interesting fact (Place Names of Lancashire, p. 32). Examples are : Boughton < *Buktūn*; Haughton < *Hōktūn*.

§ 21. Final *l* is lost in the modern local pronunciation of Southwell [saðə]; Hucknall [hɑknə]; Watnall [wotnə]; Wysall [waisə]. On the other hand, an excrescent *d* was added to Arnold < *Earn hale*; see Dial. Gramm. § 306; Horn, § 188.

§ 22. **Norman Influence.** The inability of the Normans to pronounce *þ* caused them to substitute *t* for that sound. Initial *t* for *þ* has persisted in : Tollerton, Torworth, Toton, Treswell (qq.v.). (Zachrisson, pp. 39 sqq.)

They simplified the initial combination *sn* by dropping the *s* in Nottingham < *Snotingahām*. (Zachrisson, pp. 51, 55.)

§ 23. **Scandinavian Influence** accounts for the irregular development of O.E. *ġ* and *ċ*, and *sċ* in the following cases : Egmanton < *Eċġ-*; Lindrick < *-rīċ*; Fiskerton < *fisċere-* (?); Muskham < *Musċa-* (?).

[1] A late O.E. example of this change is *lēhtūn* " garden " (c. 950, Lindisf. Gosp.) < *lēactūn*; see N.E.D., s.v. *leighton*.

WORDS OTHER THAN PERSONAL NAMES IN THE NOTTINGHAMSHIRE PLACE-NAMES

I. WORDS OF ANGLO-SAXON ORIGIN

āc, oak-tree (Hodsock, Shireoaks)
ǣdre, spring, channel of water (Averham)
æsċ, ash-tree (Ashfield, Askham)

bæþ, bath (Bathley)
bearo, wood, forest (Bassetlaw)
beofor, beaver (Bevercotes)
beorg, hill, mountain (Flawborough)
beretūn, barley-enclosure, farmstead (Barton, Carburton)
blīðe, blithe, gentle (Blyth)
bold, *botl*, house, dwelling (Bole, Bolham, Newbold)
brād, broad (Bradmore, Broadholme)
*brand, forest land cleared by fire (Bramcote)
brōc, brook (Daybrook, Giltbrook)
bryċg, bridge (Bridgford)
bul(e), bull (Bulcote)
*bullan, to bubble (Bulwell)
bune, calamus, canna (Bunny)
burh, fortified place (Attenborough, Bilborough, Brough, Burton Joyce,
  Clareborough, Fledborough, Littleborough, Woodborough)
burne, spring, brook (Winkbourn)

cealf, calf (Calverton)
*cild, spring, fountain (Chilwell)
clǣfre, clover (Clareborough)
clæ̇ġ, clay (Clayworth, Sturton-in-the-Clay)
clif, rock, cliff (Clifton, Radcliffe, Ratcliffe, Rushcliff)
cot, house, cot, habitation of human beings and animals (1 : Bramcote,
  Coates, Cotham, Cottam, Oldcoates, Woodcotes; 2 : Bevercotes, Bulcote, Lamcote)
crumb, crooked, winding (Cromwell)
cucu, *cwic*, quick, fast, alive (Cuckney)
cumb, deep hollow or valley (Sowlkholme)
cȳ, form of *cū*, cow (Keyworth)
cyne- , royal (Kingston)

denu, valley (Saxondale)
dīc, ditch (Bycardyke, Heck Dyke)
draca, dragon (Drakeholes)
dræg, retreat, nook (Drayton)
dūn, hill (Dunham, Farndon, Headon)

ēa, ēġe, water meadow, island; river, stream (Blyth, Bunny, Cuckney, Drinsey Nook, Eaton, Greet, Idle, Lithe, Mattersey, Norney, Tilne)
*ealce, a mythological person (Awkley)
eald, old (Oldwark Spring)
ēast, east (Eastwood)

fearn, fern (Farndon)
feld, open country, as opposed to woodland ; a plain (Bassingfield, Farnsfield, Felley, Haggonfield, Hatfield, Kirkby-in-Ashfield, Mansfield, Netherfield)
fenn, mud, dirt; fen (Fenton, ? Finningley)
fiscere, fisherman (Fiskerton)
flint, rock (Flintham)
ford, ford (Basford, Bridgford, Flawford, Gateford, Hazelford, Langford, Radford, Retford, Rufford, Salterford, Shelford, Spalford, Stanford, Stapleford, Wilford)
fūl, dirty, soiled ; miry (Fulwood)

gædeling, companion in arms (Gedling)
gāra, triangular piece of land (Langar)
gāt, goat (Gotham)
graf, græf, grave, burial-place (Cotgrave, Grove ?)
grāf, grove (Grove ?)
grēne, green (Gringley ?)
grēot, sand, rubble (Girton, Greet)
gylden, golden (Giltbrook)

hæsel, hazel (Hazelford, Hesley)
hǣþ, heath (Hatfield, Headon)
hām, home ; see special article, p. 169 ; (Beckingham, Bingham, Collingham, Dunham, Flintham, Gotham, Hoveringham, Lowdham, Markham, Marnham, Muskham, Nottingham, Walkeringham)
healh, nook, valley ; see special article, p. 169 ; (Arnold, Cossal, Hallam, Halloughton, Hucknall, Kersall, Kneesall, Nuthall, Ordsall)
hege, haga, hedge, fence ; a piece of enclosed ground (Bilhagh, Haywood Oaks, Thorney, Wellow)
held, slope, declivity (Merrils Bridge)
here, army (Harwell ?)
hierde, Anglian *heorde*, shepherd (Harby)
hlāw, hill, mound (Bassetlaw)

*hlūd, stormy? (Lowdham)
hōh, hōg, hill, mound ; tumulus (Watnall, Wysall)
hol, hole, cave, den (Drakeholes)
hol(h), hollow (Holbeck)
holt, wood, copse (Hawton, Teversal)
*hop, valley (Styrrup, Warsop, Worksop)
hors, horse (Horsepool)
hwǣte, wheat (Whatton, Wheatley)
hyll, hill (Cropwell)
hyrst, grove, wood (Lyndhurst)

lamb, lamb (Lambley, Lamcote)
land, plough-land (Langford)
lane, lane, street (Laneham)
lang, long (Langar)
lēah, meadow, field; see special article, p. 170 ; (Annesley, Awkley, Baggalee,
    Bathley, Brinsley, Edingley, Elkesley, Felley, Finningley, Greasley,
    Gringley, Hesley, Kimberley, Lambley, Rockley, Scarle, Strelley,
    Wansley, Wheatley, Wigsley)
lind, lime-tree (Limpool, Linby, Lindrick, Lyndhurst)
līðe, lithe ; smooth, still (Lide)
lȳtel, little, small (Littleborough)

(ge)mǣr, *mǣre, boundary (Martin)
mapuldor, maple-tree (Maplebeck, Mapplewell)
mearc, boundary (Markham)
mere, lake, pool (Bradmore, Gibsmere)
middel, middle (Middlethorpe)
mōr, moor (Barnby Moor, Morton)
mylen, mill (Milnthorpe, Milton)
mynster, monastery, church (Misterton)
myrige, pleasant, agreeable, delightful (Merrils Bridge)

neoðerra, lower (Netherfield)
netele, nettle (Nettleworth)
niuwe, new (Newark, Newbold, Newington, Newstead, Newthorpe, Newton)
norð-, north (Norton, Norwell)
norðerne, northern (Norney)

oxa, ox (Oxton)

papol, pebble (Papplewick)
plūm-trēo, plum-tree (Plumptree)
pōl, pool, pond (Horsepool, Limpool, Widmerpool)

rēad, red (Radcliffe, Radford, Ratcliffe)
*rīc, wood, forest ; tract of land ? (Lindrick)

risc, rysc, rush (Rushcliff)
rūh, rough (Rufford)

sǣta, settler, dweller ; inhabitant (Bassetlaw)
*sceld, shallow (Shelford)
scīr, boundary (Sherwood)
scylfe, Anglian *scelfe,* shelf, ledge (Shelton)
sealh trēo, sallow-tree (Salterford)
spring, spring, fountain (Oldwark Spring)
stān, stone (Kingston, Stanford, Stanton, Staunton)
stapol, pillar (Stapleford)
stede, place (Newstead)
stīepel, Anglian *stēpel,* steeple (Sturton-le-Steeple)
*stoc, stocc, stump of a tree, stake ; enclosed place ; log hut ? ; stockade ?
see discussion under Stoke Bardolph ; (Costock, Stockwith, Stoke
Bardolph, East Stoke, Stokeham)
stōw, place ; holy place (Broxtow, Edwinstowe)
strǣt, street, paved road (Strelley, Sturton)
*sulh, miry place, swamp (Sowlkholme)
sūð, south (Southwell, Sutton)

trēow, tree (Plumptree, Trowell)
tūn, enclosure ; farmstead, see special article, p. 171 ; (Adbolton, Alverton,
Aslockton, Babbington, Balderton, Barnston, Barton, Beeston, Bon-
nington, Boughton, Broughton, Burton Joyce, West Burton, Calverton,
Carburton, Car Colston, Carlton, Caunton, Clifton, Clipston, Coddington,
Colston Basset, Dalington, Darlton, Drayton, Eaton, Edwalton, Egman-
ton, Elston, Elton, Epperston, Everton, Fenton, Fiskerton, Gamston,
Girton, Glapton, Gonalston, Goverton, Grimston, Halloughton, Haugh-
ton, Hawton, Hayton, Hockerton, Kilvington, Kinoulton, Kirklington,
Kirton, Kneeton, Laxton, Lenton, Leverton, Manton, Martin, Milton,
Misterton, Morton, Newington, Newton, Normanton, Norton, Ollerton,
Ompton, Orston, Osberton, Ossington, Oxton, Rampton, Rempston,
Rolleston, Ruddington, Scarrington, Scofton, Screveton, Selston, Shelton,
Sneinton, Stanton, Staunton, Sturton, Sutton, Syerston, Thoroton,
Thrumpton, Thurgarton, Tollerton, Toton, Upton, Weston, Whatton
Wimpton, Wiseton, Wiverton, Wollaton)

þorn, thornbush (Thorney)

ūle, owl (Oldcoates)
ūp, above (Upton)

weald, Anglian *wald,* forest (the Wolds)
wealh, slave, serf ; Briton ; may be a pers. n. (Walesby)
weallan, to boil ; to flow, go in waves (Wallingwells)
weorc, worc, building ; fortification (Newark, Oldwark Spring)

*weorþ, worþ, enclosed homestead ; habitation ; see special article, p. 171 ; (Awsworth, Babworth, Blidworth, Clayworth, Colsterworth, Harworth, Hawksworth, Keyworth, Littleworth, Nettleworth, Rainworth, Scaftworth)

west, west (Weston)

wiell, well, spring, fountain ; stream (Bulwell, Chilwell, Cromwell, Harwell, Mapplewell, Norwell, Southwell, Wallingwells, Welbeck, Welham, Wellow)

*wiluh, *wilug, willow-tree (Barnby-in-the-Willows, Willoughby)

wiðig, willow (Witham)

wudig, woody (Woodcotes)

wudu, wood, forest (Bestwood, Fulwood, Haywood, Sherwood, Woodborough)

II. Words of Scandinavian Origin

bekk(r), brook (Beck, Bycardyke, Holbeck, Maplebeck, Oswardbeck, Welbeck)

birki-, birch (Birkland)

breiðr, broad (Bradebusk)

brekka, brink, slope (Brecks)

busk(r), shrub ; bush (Bonbusk, Bradebusk)

bȳ(r), habitation, farm ; village (Barnby Moor, Barnby-in-the-Willows, Bilby, Bleasby, Budby, Granby, Harby, Kirkby, Linby, Ranby, Saundby, Scrooby, Serlby, Skegby, Thoresby, Tythby, Walesby, Willoughby)

dal(r), valley (Saxondale)

drengr, companion ; sergeant-at-arms, may be a pers. n. (Drinsey Nook)

drit, dirt, M.E. dritig, dirty (Dirty Hucknall, or Hucknall-under-Huthwaite)

geit, goat (Gateford)

heið(r), heath (Hayton)

hesli, hazel (Hazelford, Heseland (see Birkland), Hesley)

holm(r), island (Broadholme, Holme)

kelda, well (Bothamsall, Kelham, Ranskill)

kirkja, church (Kirkby, Kirton)

kjar(r), M.E. ker, swamp, marshy ground (Carburton, Car Colston)

kropp(r), hump (Cropwell)

lœk(r), brook, rivulet (Leake)

lund(r), wood, grove (Birkland and Heseland, Lound)

skjalf, *skelf-, shelf ledge (Ranskill, early forms)

þorþ, village, hamlet, see special article, p. 171 ; (Appesthorpe, Bagthorpe, Beesthorpe, Bilsthorpe, Caythorpe, Danethorpe, Gleadthorpe, Gold-thorpe, Grassthorpe, Gunthorpe, Knapthorpe, Middlethorpe, Milnthorpe, Newthorpe, Osmondthorpe, Owthorpe, Perlethorpe, Sibthorpe, Stay-thorpe, Stragglethorpe, Thorpe-in-Glebe, Thorpe-by-Newark, Wigthorpe, Winthorpe)

þveit, a piece of land, a single farm, a hamlet (Eastwood, Huthwaite)

vað, a wading place, ford (Langwith ?, Stockwith ?)
vík, bay, creek (Colwick, Papplewick)
við(r), tree, wood, forest (Langwith ?, Stockwith ?)
voll(r), *valðuz*, field (Trowell ?)

III. Words of French and Latin Origin

bellum (Low Latin), fair, beautiful (Beauvale)
beste, beast of the chase (Bestwood)

faba, bean (Barton-in-Fabis)
forest, wood not enclosed, forest (Lyndhurst-on-the-Forest)

glebe, Latin *glēba*, plough-land (Thorpe-in-Glebe)
grange, granary ; outlying farm-house (Gleadthorpe Grange)

vallum, vale, valley (Beauvale)

APPENDIX

SOME OF THE MORE FREQUENT SUFFIXES EXPLAINED

(1) *hām*. The meaning is clear: "home, house, abode, estate." It denotes the dwelling of some person of consequence, or the chief seat of a tribe or noble family. Bede renders the pl. n. *Rendlæsham* by "mansio Rendili," where *mansio* stands in its Low Latin sense, from which both English *mansion* and French *maison* are derived.

(2) *healh*, dative *hēale, hāle, hale*. Although this termination is of frequent occurrence in pl. ns., it is extremely difficult to give its exact signification. After careful consideration of the evidence, Professor Wyld arrives at the conclusion that it means "a hollowed-out area, a bay, retreat" (Place-Names of Lancashire, p. 340). Miller (Place-Names in Bede, pp. 38, 39) discusses this word at length ; he translates it by "recess, corner, hollow." I am inclined to go further than that and assign the meaning of "valley" to O.E. *healh*. This view is supported by geographical evidence : all the places containing this element seem to lie in a "hollow," or, at any rate, to be situated

close to a valley where the original habitation may have stood. Bede's translation of *Streones halh* as "Sinus fari" has been a puzzle to many writers, and I do not propose to solve the mystery of the first element. Whether *farus* means "light-house" or not is a question that may perhaps never be decided. The signification of the second element is clear : *sinus* means, of course, "bay," but it has another sense as well, in which it is used here, namely that of "mediterraneus terrae angulus," as e.g. in the combination "vallium sinus." (Forcellini, Totius Latinitatis Lexicon, Schneeberg, 1831.) It describes a triangular piece of land forming the bottom of a valley between two hills or ranges that meet at one end. Such "corners, nooks, retreats, inland bays" or whatever the description may be, must have been the very spots to attract the early settlers. For they were more easily cleared of trees and undergrowth, if there were any, than the hill-sides; they were almost invariably watered by a small stream, and afforded shelter and protection.

As to the derivation of the word *healh*, I am inclined to connect it by gradation with O.E. *holh*, "hollow." The modern Frisian *hallich*, pl. *halligen*, I regard as identical[1]. Dijkstra (Friesch Woordenboek, Leeuwarden, 1900) gives its meaning as follows : "Kleine anbeduinde en onbedijkte eilandje aan de Noordfriesche kust,—overblijpselen van door de see verzwolgen land, waar de bewoners nog Friesch spreken." One usually connects the latter word with O.E. *hyll* etc., being under the impression that it refers to the artificial mounds, or "Werften," on which the houses of the "Halligen" are erected. This, however, is not correct. Originally, the term was applied to low-lying land not protected from the sea by dykes, and therefore subject to being flooded. I am told that the latter meaning is still the one attached to the word *hallich* by the islanders themselves.

(3) *lēah*, Mercian *lēh*, dative *lēge*. This word is related to Latin *lucus*, "grove," and O.H.G. *lôh*, "brushwood, clearing"; the meaning of modern Germ. *Loh* is "grove, copse." The original sense is that of "clearing, open space in the wood." The development of senses in the German and Latin words is easily explained and affords another example of how completely the meaning of geographical terms may change. The *lucus* or clearing does not consist of the open space only, but it comprises also the surrounding trees. In course of time, the latter came to be regarded as the essential and characteristic feature, the word thus assuming the meaning of "collection of trees, grove."

In the modern dialects, *lea* is used in the following senses : "meadow, field, pasture, grassland." In the pl. ns., I have translated it by "field," which may refer both to grazing and arable land. In O.E. charters, *lēah* is

---

[1] The development of a svarabhakti vowel between *l* and *h* is a regular feature of the Frisian language; see Siebs, "Geschichte der friesischen Sprache," § 85, in Pauls Grundriss, I[2]. Prof. Siebs kindly informed me that, as far as sound-development is concerned, *hallich* might very well be identical with O.E. *healh*.

rendered by Latin *campus* (see N.E.D., s.v.); Asser (Life of King Alfred, ed. Stevenson) renders *Aclea* by "in campulo quercus."

(4) *tūn*. This word is identical with O.H.G. *zûn*, modern Germ. *Zaun*, hedge, fence." Thus it originally referred to the paling or hedge with which the Teuton settler surrounded his homestead in order to protect himself and his beasts from the attacks of wild animals as well as of thieves. This original sense of the word appears clearly in the O.E. compound *dēor-tūn*, "a deer enclosure, a frith, a park." Its application, however, became soon extended. It was used to denote not only the actual palisade or hedge, but everything inclosed by the latter, namely, the whole homestead or farm. That is still the meaning of the Scotch *toon* which the Dial. Dict. explains as "farmstead, farmhouse and buildings, country seat, single dwelling." In the latter senses, the suffix was used in the pl. ns., which accounts for the fact of so many of the names in *-tūn* having a pers. n. for their first element. The modern meaning of the independent word *town* is a late development.

(5) *þorp*. Although it is a fact that this word is found in O.E. before the Scandinavian invasion, I am inclined to regard it as of Norse origin when it occurs as the second element of pl. ns. Professor Wyld holds the view that it may be derived from either source. It makes its most frequent appearance in districts containing a large Scandinavian population, and is very often compounded with Norse pers. ns. Professor Moorman (Place-Names of the West Riding, p. xlv) points out that it is very common as a termination in Danish pl. ns., but is more rarely encountered in Norway and hardly at all in Iceland. This seems to explain the fact that there are but few examples of its occurrence in Lancashire where the Scandinavian element is largely of Norwegian descent, whereas it is much more frequent in the Danish districts of Yorkshire and Notts.

The meaning of *þorp* is "village." Originally it described a collection of dwellings, and is thus opposed in signification to the single and more imposing habitations called *hām*, *tūn*, and *weorþ*. Whereas the latter were owned by a person of consequence or occupied by a noble family whose "hall" or "seat" they were, the former seems to have been composed of the more humble cots and huts of serfs, common people, or the great man's retainers. It is noteworthy, however, that *þorp* is frequently preceded by a pers. n. in the singular; it is doubtful whether the latter referred to the lord and owner of the village, or whether the word *þorp* had lost its primitive meaning and become identical with *tūn* and *hām* denoting single dwellings.

(6) *weorþ*, *worþ*. The O.E. pl. n. *Beodricesweorð* is translated into Latin by "Bedrici curtis" (Passio Sancti Edmundi, c. 14). The meaning of Low Latin *curtis* is "enclosure, estate." Both the original meaning and the etymology of O.E. *worþ* are obscure. Professor Skeat (Place-Names of Cambridgeshire, p. 25) connects it with O.E. *weorþ*, "worth, value," which does not seem a very happy explanation. I venture to suggest that M.H.G. *wert, -des*,

"island, peninsula, raised dry land between morasses," is a possible cognate. Dr Hirt (Weigands Deutsches Wörterbuch, s.v. *Werder*) derives *wert* from the root contained in Gothic *warian*, O.E. *werian*, "to protect, ward off; dam up." If that be correct, *weorþ* would originally have been applied to a piece of land—with or without a dwelling—protected by a dam or dyke, or possibly a palisade. The transition from this primary sense to that of "farmstead, habitation, estate" is natural and parallel to that observed in the history of the suffix *tūn*.

The relation between the two German forms *werd* and *werder* is the same as between O.E. *sæl—salor*, "hall," *sige—sigor*, "victory" (Sievers, § 288, 289).

THE PRINCIPAL PERSONAL NAMES IN THE
NOTTINGHAMSHIRE PLACE-NAMES

I. ANGLO-SAXON AND NORSE PERSONAL NAMES

Adda, Adding (Attenborough)
Æȝel or *Æðel* (Elton)
Ælf(a) (Elston)
Ælfhere (Alverton, Ollerton)
? *Alca* (Awkley)
Anna (Annesley)
Āslakr (Aslockton)

Babba (Babbington, Babworth)
? *Bada* (Bathley)
Bagga (Baggalee, Bagthorpe)
Barn (Barnby)
Basing (Basingfield)
Bassa (Basford)
Bealdhere (Balderton)
Becca, Beccing (Beckingham)
Benna, Benning (Bingham)
Beorn (Barnston)
? *Bildi* (Bilsthorpe)
? *Bilheard* (Bilsthorpe)
Billa (Bilborough, Bilby)
? *Bliðhere* (Blidworth)
*Bodmǣr or *Bodwine* (Bothamsall)
Bondi (Bonbusk)
Bonningas (Bonnington)
*Brōcwulf (Broxtow)
Brȳn, Brūn (Brinsley)
Bucea (Boughton)
Būtr or *Butti* (Budby)

? *Carl* (Carlton, Caythorpe)
Clip (Clipston)
? *Cniht* (Kneesall, Kneeton)
Codda, Codding, Cotta (Coddington)
Col, Colla, Colling (Car Colston, Collingham, Colston Basset, Colwick)

Cortel, Corteling (Costock)
Cotta (Cossal)
Cylfa (Kilvington)
Cynemǣr (Kimberley)
Cyneweald (Kinoulton)
**Cyrtel* (Kirklington)

Dēorlāf (Darlton)
Dēorling (Dalington)
Deorna (Danethorpe)

Eada (Edingley ?)
Eadbeald or *Ealdbeald* (Adbolton)
? *Eadwæcer* (Eakring)
Eadweald (Edwalton)
Eadwine (Edwinstowe, Edingley ?)
Ealda (Awsworth)
Ealhmund (Ompton)
Earne (Arnold)
Ecgmund (Egmanton)
**Elc* (Elkesley)
Eobeorht, **Eoperht* (Epperston)
Eofor, Eoforing (Everton, Hoveringham)

Flǣda (Fledborough)
? *Fræna* (Farnsfield)

Gamal (Gamston)
Gārfriđ (Goverton)
Glæppa (Glapton)
Golda (Goldthorpe)
Grani (Granby)
Grīmr (Grimston)
Grīs (Greasley)
Gunner or *Gunnild* (Gunthorpe)
Gunnulf (Gonalston)
Gybba, Gyppa (Gibsmere)

Haukr, Hōc (Hawksworth, Hockerton)
Heara (Harworth)
Hiccelingas (Hickling)
Hod or *Oddi*? (Hodsock)
Hrafn (Rampton, Ranby, Ranskill)
Hrefn (Rempston)
Hrōaldr, Rold (Rolleston)
**Hucca* (Hucknall, Huthwaite)

*Laxa, Leaxa (Laxton)
? Lēofhere (Leverton)

Mǣna (Manton)
Mǣrwingas (Meering)
Mǣ∂here (Mattersey)
*Mamma (Mansfield)

Ordrīc (Ordsall, Orston)
Ōsbeorn, Āsbeorn (Osberton)
*Ōskin, Āsketill (Ossington)
Ōsmund, Āsmund (Osmondthorpe)
Ōsweard or Ōsweald (Oswardbeck)

? Ragnald, Regnald or Ragnhildr (Ragnall)
Ruddingas (Ruddington)

? Saxi or Seaxa (Saxondale)
Selfa (Selston)
Serlo (Serlby)
Sibbi (Sibthorpe)
Sigerič (Syerston)
Skarf or Sceorf (Scarle, Scarrington?)
? Skar∂i, Skar∂ingas (Scarrington?)
Skegge (Skegby)
Skopti (Scofton)
*Snottingas (Nottingham, Sneinton)
Steorra, Styr (Staythorpe, Styrrup)
? Strangwulf (Stragglethorpe)

Tila (Tilne)
Tucca (Tuxford)

þor (Thoresby)
þorleifr, þorleikr (Tollerton)
þor∂r, þure∂ (Torworth)
þurfer∂ (Thoroton)
þurgeir, þurgār (Thurgarton)
þurmō∂r (Thrumpton)
þurulf (Toton)
? þyri (Treswell)

Ūfi (Owthorpe)

*Wanda (Wansley)
*Wara (Warsop)
Wata (Watnall)
? Wealh (Walesby)
Wealhhere, Walchere (Walkeringham)

Wĭdmǣr (Widmerpool)
Wigfrĭð, Wigferð or *Wĭdfara*? (Wiverton)
? *Vīgi* (Wigsley, Wigthorpe)
Willa, Will (Wilford)
Winemund (Wimpton, Winthorpe)
*Wīsa (Wiseton, Wysall)
Wulflāf (Wollaton)
**Wyre,* **Weorc,* *Worc (Worksop)

II. NORMAN-FRENCH PERSONAL NAMES

Bardolf (Stoke Bardolph)
Basset[1] (Colston Basset)
Butler (Cropwell Butler)
Cantelupe (Watnall Cantelupe)
Joyce, older *Jorze* (Burton Joyce)
Peverel (Perlethorpe)
Pierrepont (Holme Pierrepont)
Torkard (Hucknall Torkard)

N.B. Several of the above names are explained in the index to the Calendar of Documents Preserved in France, edited by J. Horace Round, 1899.

[1] This name was adopted by the family in England, being taken from the Nottinghamshire Hundred name of Bassetlaw (q.v.).

BIBLIOGRAPHY

I. SOURCES OF EARLY FORMS OF NOTTINGHAMSHIRE
PLACE-NAMES.

A.S. CHRON. Anglo-Saxon Chronicle.
BEDE, HIST. ECCLES. Historia Ecclesiastica Gentis Anglorum.
BODL. CH. & R. Calendar of Charters and Rolls Preserved in the Bodleian Library. Oxford, 1878.
BOR. REC. Records of the Borough of Nottingham. Vol. I. 1882.
CAL. ROT. CHART. Calendarium Rotulorum Chartarum, etc. (Record Office). 1803.
CAMDEN. Britain, or a Chorographicall Description of the Most flourishing Kingdoms, England, Scotland, and Ireland....Written by William Camden. 1637.
CART. SAX. Cartularium Saxonicum, ed. Birch.
COD. DIPL. Codex Diplomaticus Ævi Saxonici, ed. Kemble.
D.B. Doomsday Book (in Victoria County History, Vol. I).
DRAYTON. Drayton's Polyolbion (modernised).
E.E.T. SOC. Early English Text Society, Publications.
F.A. Inquisitions and Assessments relating to Feudal Aids (Record Office). 1899 etc.
FOR. REC. Forest Records, edited by W. H. Stevenson.
H.R. Rotuli Hundredorum (Record Office). 1812–18.
INDEX. Index to Charters and Rolls in the British Museum. Vol. I. Index Locorum. 1900.
INQ. P.M. Calendar of Inquisitions Post Mortem (Record Office), Vol. I. 1904 ; Vol. II. 1906.
INQ. P.M. c. 1500. Abstracts of the Inquisitions P.M. Vol. I. 1485–1546 (Thoroton Society, Records Series III). 1905.
LELAND. The Itinerary of John Leland the Antiquary. 2nd edition. Oxford, 1745.
MAP IN CAMDEN. See Camden, above.
MAP 1704. Map of Nottinghamshire by R. Morden. 1704.
MON. ANGLIC. Dugdale, Monasticon Anglicanum. New edition.
NON. INQ. Nonarum Inquisitiones in Curia Scaccarii. Temp. Regis Edw. III (Record Office). 1807.
NOTTM. CH. Royal Charters granted to the Burgesses of Nottingham. 1890.

P.R. The Great Rolls of the Pipe (Pipe Roll Society). 1884 etc.
TAX. ECCLES. Taxatio Ecclesiastica Angliae et Walliae (Record Office). 1802.
TESTA DE N. Testa de Nevil sive Liber Feodorum (Record Office). 1807.
THOROTON. Thoroton's History of Nottinghamshire. Republished with Large Additions, by John Throsby. 1797.
VALOR ECCLES. Valor Ecclesiasticus. Temp. Regis Hen. VIII. Vol. v (Record Office). 1825–34.
WOLL. MSS. Report on the Manuscripts of Lord Middleton Preserved at Wollaton Hall, ed. W. H. Stevenson (Hist. MSS. Commission). 1911.

II. OTHER BOOKS RELATING TO NOTTINGHAMSHIRE.

GUILDFORD, E. L. Nottinghamshire (The Little Guides Series).
WHITE, F. History, Directory and Gazetteer of the County, and the Town and County of the Town of Nottingham. Sheffield, 1853.
WHITE, ROBERT. The Dukery Records. 1904.

III. SOURCES OF PERSONAL NAMES.

SEARLE. Onomasticon Anglo-Saxonicum. 1897.
BJÖRKMAN. Nordische Personennamen in England. Halle, 1910.
RYGH. Gamle Personnavne i norske Stedsnavne. Kristiania, 1901.
RYGH. Norske Gaardnavne, I. Kristiania, 1897.
BARDSLEY. A Dictionary of English and Welsh Surnames. 1901.
FÖRSTEMANN. Altdeutsches Namenbuch, I. Personennamen. Bonn, 1900.
WINKLER. Friesche Naamlijst (Onomasticon Frisicum). Leeuwarden, 1898.
WERLE. Die ältesten germanischen Personennamen. Strassburg, 1910.
SOCIN. Mittelhochdeutsches Namenbuch. Basel, 1903.

IV. MONOGRAPHS ON PLACE-NAMES.

ALEXANDER. Place-Names of Oxfordshire. 1912.
ALEXANDER. The Particle -ing in Place-Names. (Essays and Studies by Members of the English Association, Vol. II.)
ALEXANDER. The Genitive Suffix in the First Element of English Place-Names. (Modern Language Review, 1911.)
BRADLEY, HENRY. English Place-Names. (Essays and Studies by Members of the English Association, Vol. I.)
DUIGNAN. Place-Names of Staffordshire ;— Warwickshire ;— Worcestershire.
FÖRSTEMANN. Altdeutsches Namenbuch, II. Ortsnamen. Nordhausen, 1872.
JELLINGHAUS. Englische und niederdeutsche Ortsnamen. (Anglia, XX. pp. 257–334.)
MACCLURE. British Place-Names in their Historical Setting. 1910.

MIDDENDORF. Altenglisches Flurnamenbuch. Halle, 1902.
MILLER. Place-Names in the Old English Bede. Strassburg, 1896.
MOORMAN. Place-Names of the West Riding of Yorkshire. (Transactions of the Thoresby Society.) Leeds, 1910.
NAPIER AND STEVENSON. The Crawford Collection of Charters. 1895.
SCHRÖDER, EDWARD. Über Ortsnamenforschung. (Harzverein für Geschichte und Altertumskunde.) Wernigerode, 1908.
SEPHTON. Handbook of Lancashire Place-Names. Liverpool, 1913.
SKEAT. Place-Names of Bedfordshire ;—Berkshire ;—Cambridgeshire ;—Hertfordshire ;—Huntingdonshire.
STURMFELS. Die Ortsnamen Hessens. Leipzig, 1910.
TAYLOR. Words and Places.—English Village Names. (Reprinted in Everyman's Library.)
WYLD AND HIRST. Place-Names of Lancashire. 1911.
ZACHRISSON. A Contribution to the Study of Anglo-Norman Influence on English Place-Names. Lund, 1909.

V. GENERAL BOOKS OF REFERENCE ON GRAMMAR, ETYMOLOGY, AND PHONOLOGY.

BOSWORTH-TOLLER. An Anglo-Saxon Dictionary.
BÜLBRING. Altenglisches Elementarbuch, I. Heidelberg, 1902.
HIRT. Weygandt's Deutsches Wörterbuch.
HORN. Historische Neuenglische Grammatik, I. Strassburg, 1908.
KLUGE. Etymologisches Wörterbuch der deutschen Sprache.
MORSBACH. Mittelenglische Grammatik, I.
MURRAY, BRADLEY, AND CRAIGIE. New English Dictionary [N.E.D.].
SIEVERS. Angelsächsische Grammatik. 3rd edition. Halle, 1898.
SKEAT. Etymological Dictionary of the English Language. 4th edition.
STOLZE. Zur Lautlehre der altenglischen Ortsnamen im Domesday Book. Berlin, 1902.
STRATMANN-BRADLEY. Middle English Dictionary. 1891.
SWEET. Student's Dictionary of Anglo-Saxon. 1897.
SWEET. New English Grammar, Logical and Historical. 1900.
VIGFUSSON-CLEASBY. Icelandic Dictionary. 1874. [Vigf.]
WRIGHT. English Dialect Dictionary [Dial. Dict.].
WRIGHT. English Dialect Grammar [Dial. Gramm.].